Lecture Notes in Mathematics

Editors:
A. Dold, Heidelberg
B. Eckmann, Zürich
F. Takens, Groningen

Eberhard Freitag

Singular Modular Forms and Theta Relations

Springer-Verlag

Berlin Heidelberg New York
London Paris Tokyo
Hong Kong Barcelona
Budapest

Author

Eberhard Freitag
Mathematisches Institut
Universität Heidelberg
Im Neuenheimer Feld 288
W-6900 Heidelberg, FRG

.

Mathematics Subject Classification (1991): 11F, 10D

ISBN 3-540-54704-5 Springer-Verlag Berlin Heidelberg New York
ISBN 0-387-54704-5 Springer-Verlag New York Berlin Heidelberg

© Springer-Verlag Berlin Heidelberg 1991
Printed in Germany

Typesetting: Camera ready by author
Printing and binding: Druckhaus Beltz, Hemsbach/Bergstr.
46/3140-543210 - Printed on acid-free paper

Eberhard Freitag

Singular Modular Forms and Theta Relations

Springer-Verlag

Berlin Heidelberg New York
London Paris Tokyo
Hong Kong Barcelona
Budapest

Author

Eberhard Freitag
Mathematisches Institut
Universität Heidelberg
Im Neuenheimer Feld 288
W-6900 Heidelberg, FRG

Mathematics Subject Classification (1991): 11F, 10D

ISBN 3-540-54704-5 Springer-Verlag Berlin Heidelberg New York
ISBN 0-387-54704-5 Springer-Verlag New York Berlin Heidelberg

© Springer-Verlag Berlin Heidelberg 1991
Printed in Germany

Typesetting: Camera ready by author
Printing and binding: Druckhaus Beltz, Hemsbach/Bergstr.
46/3140-543210 - Printed on acid-free paper

Table of contents

Table of contents

Introduction

Siegel modular forms are holomorphic functions on the generalized upper half plane $\mathbb{H}_n$, which consists of all symmetric complex $n \times n$-matrices with positive definite imaginary part. They admit Fourier expansions of the type

$$f(Z) = \sum_T a(T) \exp \pi i \sigma(TZ),$$

where T runs through a lattice of rational symmetric matrices.

A modular form is called **singular** if

$$a(T) \neq 0 \Longrightarrow T \quad \text{singular, i.e.} \quad \det T = 0.$$

Important examples of modular forms are **theta series**

$$\sum_G \exp \pi i \sigma(S[G]Z).$$

Here S is a rational symmetric positive definite $r \times r$-matrix and G runs through a lattice of rational $r \times n$-matrices. This theta series is singular if

$$r < n.$$

The theory of singular modular forms states in a very precise sense that each singular modular form is a linear combination of theta series.

In these notes we give an introduction to the theory of **Siegel modular forms**, especially singular ones. We aspire to highest generality, we consider arbitrary congruence subgroups of the Siegel modular group, the modular forms can be vector valued and the weights are allowed to be half integral. Before we describe the contents in more detail, we make some historical comments.

The notion of a singular modular form is due to H.RESNIKOFF [Re1], who considered scalar valued modular forms of the transformation type

$$f(M\langle Z \rangle) = v(M) \det(CZ + D)^{r/2} f(Z).$$

He proved that such a modular form can be singular only if r is integral and if $r < n$. The question arose whether in this case all modular forms are singular. In the paper [Fr1] an affirmative answer was given in the case $n = 2$. A little later the general case was independently solved by different methods in [Fr2] and [Re2].

The only known examples of singular modular forms were theta series (and linear combinations of them). So the question arose whether every singular modular form is a linear combination of theta series. I gave a very short and simple proof (s.[Fr4]) that scalar valued singular modular forms with respect to the full Siegel modular group are linear combinations of theta series

$$\sum_{G \text{ integral}} \exp \pi i \sigma(S[G]Z),$$

where S is an even unimodular positive matrix. (Even means that S is integral and that the diagonal is even.) A little later I generalized this result to vector valued modular forms with respect to the full Siegel modular group [Fr5]. Those forms are holomorphic functions

$$f : \mathbb{H}_n \longrightarrow \mathcal{Z}$$

with values in some finite dimensional vector space $\mathcal{Z}$, which transform as

$$f(M\langle Z \rangle) = \varrho(CZ + D)f(Z).$$

Here

$$\varrho : \mathrm{GL}(n, \mathbb{C}) \longrightarrow \mathrm{GL}(\mathcal{Z})$$

is some rational representation. The role of r is played by the biggest number k such that

$$\varrho(A) \det(A)^{-k}$$

is polynomial. In the vector valued case it is necessary to consider theta series with harmonic coefficients:

$$\sum_G P(S^{1/2}G) \exp \pi i \sigma(S[G]Z).$$

Here P is a harmonic polynomial with the transformation property

$$P(GA) = \varrho(A') \det(A)^{-r/2} P(G).$$

(For simplicity we assume $r \equiv 0 \bmod 2$ at the moment.) It is a very remarkable fact that vector valued singular modular forms automatically "produce" such harmonic coefficients.

The method which I used depended heavily on the restriction to the full modular group. As has been pointed out in [Al], [Zw] and [En], the method generalizes to congruence groups which contain all unimodular substitutions,

$$Z \longmapsto Z[U], \quad U \in \mathrm{GL}(n, \mathbb{Z}).$$

But the general case of an arbitrary congruence subgroup or at least of a cofinal system of congruence subgroups was not obvious.

In an important paper [Ho] R.Howe proved a theorem about singular representations of $\mathrm{Sp}(n, \mathbf{A})$ ($\mathbf{A}$ denotes the ring of adeles). In classical language his result can be formulated as follows:

Each singular Siegel modular form is linear combination of theta series.

(HOWE actually considered only square integrable modular forms with respect to the PETERSSON inner product. A little later WEISSAUER proved [Wel] that singular forms are always square integrable.).

In some sense the result of HOWE is not satisfactory. Let Γ be a fixed congruence subgroup of the Siegel modular group. As usual we denote by $[\Gamma, \varrho]$ the space of all modular forms with the transformation law

$$f(M\langle Z \rangle) = \varrho(CZ + D)f(Z) \quad \text{for all } M \in \Gamma.$$

This space is of finite dimension. Very often the dimension has geometric or arithmetic meaning and one would like to compute or estimate it.

*For this reason one would like to have a **finite system of theta series** which generates $[\Gamma, \varrho]$ and one would like to describe all linear relations between the generators.*

The representation theoretic result of HOWE seems not to give an answer to this refined question. For example, he always has to consider besides the theta series

$$\sum_{G \text{ integral}} P(S^{1/2}G) \exp \pi i \sigma(S[G]Z)$$

all satellites

$$\sum_{G \text{ integral}} P(S^{1/2}G) \exp \pi i \sigma(S[G]Z + 2G'V)$$

with arbitrary rational characteristics V. But they generate a vector space of infinite dimension and it is not clear which linear combinations of them belong to a given group Γ. This question is difficult because there are non-trivial relations between them. For example, the **classical Riemann theta relations** are of this type.

The first step into a more concrete representation theorem has been done by R.ENDRES [En]. He treated the case of scalar valued modular forms of weight $1/2$ ($r = 1$). Some of his ideas have proved to be essential for the general case.

In these notes we prove a refined representation theorem for singular modular forms, which gives a finite system of generators and describes all linear relations between them.

Actually the proof is complete only for $n \geq 2r$ (instead of $n > r$) and some other cases.

Our method is elementary and not representation theoretic. It depends heavily on the analysis of the Fourier-Jacobi expansion of a modular form. Now we describe the contents in more detail.

Chapter I contains an introduction to the theory of Siegel modular forms. We consider vector valued forms and also admit half integral weights. Therefore we have to deal with multiplier systems. The choice of a multiplier system is not too important, because in the case $n > 1$, two multiplier systems always agree on a suitable congruence subgroup. We investigate the standard multiplier system –the so called **theta multiplier system**– in some detail and express it as a Gauss sum. This Gauss sum will be computed in important special cases. All the results about the theta multiplier system are already in the literature but are scattered. It seemed me to be worth while to include this theory with complete proofs.

As already mentioned, vector valued modular forms involve finite dimensional rational representations of $GL(n)$. We include the theory of highest weights of such representations without proofs. Readers who are interested only in scalar valued modular forms can pass over this part.

Chapter II is devoted to the **theta transformation formalism**. Examples of modular forms are theta series. In the vector valued case (and only in this case) one needs theta series with harmonic coefficients. For our purposes it is necessary to develop the transformation formalism for arbitrary polynomial coefficients. One reason is that in the Fourier Jacobi expansion of a vector valued modular form, theta series arise with polynomial coefficients which are not known to be harmonic in advance. For the proof of the transformation formalism we use EICHLER's imbedding trick. This is a very convenient method to reduce the transformation formalism to the full modular group, where simple generators are available. The transformation formalism simplifies considerably if the coefficients are harmonic forms. For the purposes of these notes not much more than the definition of a harmonic form is needed. Nevertheless we have included some of the beautiful results of KASHIWARA-VERGNE [KV], who classified all harmonic forms.

Chapter III contains the proof that non-vanishing modular forms are singular if and only if $r < n$. The main tool is the **Fourier Jacobi expansion** of a modular form. The transformation properties of those coefficients lead to the notion of a **Jacobi form**. We prove a variant of the SHIMURA-isomorphism [Sh], which states that Jacobi forms correspond to finite systems of usual modular forms. The proof of this correspondence is tedious in the vector valued case and depends heavily on the general theta transformation formalism. We prove the correspondence between Jacobi forms and usual modular forms only for varying levels, i.e., we do not get information about precise levels.

Chapter IV describes a central part of the theory. First of all we describe a certain space of Fourier series M, which contains the space of modular forms of a fixed level q. This inclusion is nothing else but a reformulation of the classification of singular weights. The space M has the advantage that in its definition no multiplier system or sophisticated congruence groups have to be considered. At first glance the space M looks tremendously big. Actually our general representation theorem is valid for arbitrary elements of M. We will investigate the Fourier Jacobi expansion of elements of M. The representation theorem will be reduced to an elementary statement (called the fundamental lemma in these notes) which has nothing to do with modular forms. Unfortunately this lemma seems to be very hard.

Chapter V is devoted to the fundamental lemma. We give a complete proof in the case $n \geq 2r$ and in some other cases.

In the last chapter we formulate the results and point out the connection with the theory of theta relations. We work out a formula for the dimension of M (and as a consequence of certain spaces of singular modular forms), which allows one in principle to compute the dimension explicitly by a calculator. We include some numerical results.

But we confess that the connection between our results and the classical theta relations is not understood satisfactorily.

Most of the material of these notes has been published in the preprint series of the "Forschungsschwerpunkt Geometrie, Heidelberg" [Fr7-10].

In particular I would like to thank Dr. Dipendra Prasad who brought my attention to various mistakes in the original manuscript.

Heidelberg, 1990

I Siegel modular forms

1 The symplectic group

We introduce the symplectic group and recall some of its basic properties. A detailed treatment can be found in [Fr4].

Let R be a commutative ring with unit element $1 = 1_R$. We denote by

$$E = E^{(n)} = \begin{pmatrix} 1 & & \\ & \ddots & \\ & & 1 \end{pmatrix}$$

the $n \times n$-unit matrix with coefficients in R and by

$$I = \begin{pmatrix} 0 & E \\ -E & 0 \end{pmatrix}$$

the standard alternating matrix. The symplectic group of degree n with coefficients in R consists of all $2n \times 2n$-matrices

$$M \in R^{(2n,2n)},$$

such that

$$I[M] = I.$$

Here we use the usual notation

$$A[B] = B'AB \quad (B' = \text{transpose of } B)$$

for matrices $A \in R^{(n,n)}$, $B \in R^{(n,m)}$. We denote the symplectic group by

$$\mathrm{Sp}(n, R).$$

It is often useful to decompose a symplectic matrix into four $n \times n$-blocs:

$$M = \begin{pmatrix} A & B \\ C & D \end{pmatrix}.$$

1.1 Remark. *1) A matrix* $M = \begin{pmatrix} A & B \\ C & D \end{pmatrix}$ *is symplectic if and only if the relations*

$$A'D - C'B = E, \quad A'C = C'A, \quad B'D = D'B$$

hold. Especially

$$\text{Sp}(1, R) = \text{SL}(2, R).$$

2) One has $I^{-1} = -I$. Therefore the transpose M' of a symplectic matrix M is symplectic, i.e.

$$AD' - BC' = E, \quad AB' = BA', \quad CD' = DC'.$$

3) The inverse of a symplectic matrix is

$$M^{-1} = I^{-1}M'I = \begin{pmatrix} D' & -B' \\ -C' & A' \end{pmatrix}.$$

4) Some examples of symplectic matrices are

$$\text{a)} \quad \begin{pmatrix} E & S \\ 0 & E \end{pmatrix}, \quad S = S';$$

$$\text{b)} \quad \begin{pmatrix} U' & 0 \\ 0 & U^{-1} \end{pmatrix}, \quad U \in \text{GL}(n, R);$$

$$\text{c)} \quad I = \begin{pmatrix} 0 & E \\ -E & 0 \end{pmatrix}.$$

1.2 Proposition. *Let R be either $\mathbb{Z}$ or a field. The group $\text{Sp}(n, R)$ is generated by the special matrices*

$$\begin{pmatrix} E & S \\ 0 & E \end{pmatrix}, S = S'; \quad \begin{pmatrix} 0 & E \\ -E & 0 \end{pmatrix}.$$

The symplectic group with coefficients in $\mathbb{Z}$ is sometimes called the Siegel modular group. We denote it by

$$\Gamma_n = \text{Sp}(n, \mathbb{Z}).$$

Congruence subgroups

The kernel of the natural restriction homomorphism mod q

$$\Gamma_n[q] := \ker\bigl(\mathrm{Sp}(n,\mathbb{Z}) \longrightarrow \mathrm{Sp}(n,\mathbb{Z}/q\mathbb{Z})\bigr)$$

is called the **principal congruence subgroup** of level q.

1.3 Definition. *A subgroup*

$$\Gamma \subset \mathrm{Sp}(n,\mathbb{R})$$

is called a congruence subgroup if it contains some principal congruence subgroup

$$\Gamma_n[q] \subset \Gamma$$

as a subgroup of finite index.

1.4 Theorem. *Assume $n > 1$. Let*

$$\Gamma \subset \mathrm{Sp}(n,\mathbb{Z})$$

be a normal subgroup which is not contained in the central subgroup $\{\pm E^{(2n)}\}$. Then Γ is a congruence subgroup.

Corollary. *Each subgroup $\Gamma \subset \mathrm{Sp}(n,\mathbb{Z})$ of finite index is a congruence subgroup.*

We don't have to make use of this beautiful result of MENNICKE [Me]. But it will sometimes be helpful to have it in mind.

There are several "standard" congruence subgroups which we will use in these notes.

1) *The **theta group***

$$\Gamma_{n,\vartheta} = \{M \in \Gamma_n, \quad AB' \text{ and } CD' \text{ have even diagonal entries}\}.$$

We will see later that $\Gamma_{n,\vartheta}$ actually is a group.

2) *The generalized **Hecke group***

$$\Gamma_{n,0}[q] := \{M \in \Gamma_n, \quad C \equiv 0 \bmod q\}.$$

3) *The "theta variant" of 2)* [En]

$$\Gamma_{n,0,\vartheta}[q] = \{M \in \Gamma_n; \quad C \equiv 0 \bmod q, \text{ the diagonal entries of } (CD')/q \text{ are even}\}.$$

4) IGUSA's group [Ig1]

$$\Gamma_n[q, 2q] = \{M \in \Gamma_n[q], \quad \text{the diagonal entries of } AB'/q \text{ and } CD'/q \text{ are even}\}.$$

One has

$$\Gamma_n[2q] \subset \Gamma_n[q, 2q] \subset \Gamma_n[q].$$

Obviously

$$\Gamma_n = \Gamma_n[1]; \quad \Gamma_{n,\vartheta} = \Gamma_n[1, 2].$$

1.5 Remark. *The Igusa group*

$$\Gamma[q, 2q] \subset \Gamma_{n,\vartheta}$$

is a normal subgroup of $\Gamma_{n,\vartheta}$.
For even q, the group $\Gamma_n[q, 2q]$ is normal in the full modular group Γ_n.

The proof of this remark can be found in [Ig1].

2   The Siegel upper half space

We introduce the action of the real symplectic group on the Siegel upper half space.

In the following, we denote by

$$\mathcal{Z}_n = \{ Z = Z' \in \mathbb{C}^{(n,n)} \}$$

the vector space of all symmetric complex $n \times n$-matrices.

2.1 Definition.   *The Siegel upper half space of degree n consists of all symmetric complex $n \times n$-matrices whose imaginary part is positive (definite).*

$$\mathbb{H}_n = \{ Z = X + iY \in \mathcal{Z}_n, \quad Y > 0 \}.$$

2.2 Remark.   *The Siegel upper half space is an open convex subdomain of $\mathcal{Z}_n$.*

2.3 Remark.   *Let*

$$f : \mathbb{H}_n \longrightarrow \mathbb{C}$$

be a holomorphic function without zeros. Then there exists a holomorphic square root of f:

$$h : \mathbb{H}_n \longrightarrow \mathbb{C},$$

i.e. h is holomorphic and

$$h(Z)^2 = f(Z).$$

Proof. Consider for a fixed Z, the function

$$\alpha : [0,1] \longrightarrow \mathbb{C},$$
$$\alpha(t) = f\big(iE + t(Z - iE) \big).$$

The function

$$H(Z) := \int_0^1 \dot{\alpha}(t)/\alpha(t)\, dt$$

is holomorphic and has the property

$$e^{H(Z)} = f(Z).$$

The function

$$h(Z) = e^{H(Z)/2}$$

has the desired property.

2.4 Lemma. *Let*

$$Z \in \mathbb{H}_n, \quad M \in \mathrm{Sp}(n, \mathbb{R}).$$

Then the matrix

$$CZ + D \qquad \left(M = \begin{pmatrix} A & B \\ C & D \end{pmatrix} \right)$$

is invertible and

$$M\langle Z \rangle := (AZ + B)(CZ + D)^{-1}$$

is again contained in $\mathbb{H}_n$. *This defines an action of* $\mathrm{Sp}(n, \mathbb{R})$ *on* $\mathbb{H}_n$, *i.e.*

a) $E\langle Z \rangle = Z,$

b) $M\langle N\langle Z \rangle \rangle = (MN)\langle Z \rangle$ *for* $M, N \in \mathrm{Sp}(n, \mathbb{R})$.

The map

$$\mathbb{H}_n \longrightarrow \mathbb{H}_n,$$
$$Z \longmapsto M\langle Z \rangle,$$

is of course biholomorphic. It can be shown that each biholomorphic map of $\mathbb{H}_n$ onto itself is symplectic (i.e. of this form).

2.5 Remark. *Two symplectic matrices* $M, N \in \mathrm{Sp}(n, \mathbb{R})$ *have the same action on* $\mathbb{H}_n$ *if and only if*

$$M = \pm N.$$

Examples of symplectic substitutions are

1) $M = I = \begin{pmatrix} 0 & E \\ -E & 0 \end{pmatrix};$ $I\langle Z \rangle = -Z^{-1},$

2) $M = \begin{pmatrix} E & S \\ 0 & E \end{pmatrix}, S = S';$ $M\langle Z \rangle = Z + S,$

3) $M = \begin{pmatrix} U' & 0 \\ 0 & U^{-1} \end{pmatrix}, U \in \mathrm{GL}(n, \mathbb{R});$ $M\langle Z \rangle = U'ZU.$

For proofs we refer to [Fr4].

3 Multiplier systems

We want to include modular forms of half integral weights and not only of integral ones. This forces us to introduce the notion of a multiplier system. In the case of an integral weight a multiplier system is nothing else but a character.

3.1 Remark. *Let*

$$J(M, Z) = CZ + D \qquad (M \in \mathrm{Sp}(n, \mathbb{R}), Z \in \mathbb{H}_n).$$

Then the cocycle relation

$$J(MN, Z) = J(M, N\langle Z \rangle)J(N, Z)$$

holds.

The proof is trivial.

The function

$$Z \longmapsto \det(CZ + D)$$

posseses a holomorphic square root (2.3). This square root is determined only up to sign. We choose once and for all one of the two holomorphic square roots and denote it by

$$\sqrt{\det(CZ + D)}.$$

Warning: This notation is somewhat misleading. One has to keep in mind that the following may happen (in case $n > 1$).

There exist two points $Z_1, Z_2 \in \mathbb{H}_n$ such that

$$\det(CZ_1 + D) = \det(CZ_2 + D)$$

but

$$\sqrt{\det(CZ_1 + D)} = -\sqrt{\det(CZ_2 + D)}.$$

For an arbitrary integer r we consider

$$J_r(M, Z) = \det(CZ + D)^{r/2}.$$

From 3.1 follows

3.2 Remark. *For every $r \in \mathbb{Z}$, there exists a map*

$$w = w_r : \mathrm{Sp}(n, \mathbb{R}) \times \mathrm{Sp}(n, \mathbb{R}) \longrightarrow \{1, -1\}$$

such that

$$J_r(MN, Z) = w_r(M, N)J_r(M, N\langle Z \rangle)J_r(N, Z).$$

Of course w_r depends only on $r \bmod 2$. We have

$$w_r \equiv 1 \iff r \text{ is even.}$$

3.3 Definition. *Let*

$$\Gamma \subset \mathrm{Sp}(n, \mathbb{R})$$

be a congruence subgroup. A multiplier system of weight $r/2, r \in \mathbb{Z}$, is a map

$$v : \Gamma \longrightarrow \mathbb{C}^{\bullet} = \mathbb{C} - \{0\}$$

satisfying

a) $\qquad\qquad v(MN) = w_r(M, N)v(M)v(N) \quad \text{for all } M, N \in \Gamma.$

b) $\qquad\qquad v(-E)\det(0Z - E)^{r/2} = 1 \quad \text{if } -E \in \Gamma.$

Remark. *If v is a multiplier system, then the map*

$$J_{r,v} : \Gamma \times \mathbb{H}_n \longrightarrow \mathbb{C}^{\bullet},$$
$$(M, Z) \longmapsto v(M)\det(CZ + D)^{r/2}$$

satisfies the cocycle condition

$$J_{r,v}(MN, Z) = J_{r,v}(M, N\langle Z\rangle)J_{r,v}(N, Z).$$

Again this condition depends only on $r \bmod 2$. A multiplier system of integral weight is a character of Γ.

It is sometimes useful to use another characterization of multiplier systems:
For an arbitrary function

$$f : \mathbb{H}_n \longrightarrow \mathbb{C}^{\bullet}$$

we introduce the notation

$$f|M(Z) = f|_r M(Z) = \det(CZ + D)^{-r/2}f(M\langle Z\rangle).$$

Obviously

$$f|(MN) = w_r(M, N)(f|M)|N.$$

3.4 Remark. *A function*

$$v : \Gamma :\longrightarrow \mathbb{C}^{\bullet}$$

is a multiplier system if and only if there exists a non-vanishing function

$$f : \mathbb{H}_n \longrightarrow \mathbb{C},$$

such that

$$f|_r M = v(M)f \quad \text{for every } M \in \Gamma.$$

We describe the main example of a multiplier system of half integral weight, the so-called **theta multiplier system.**

The series

$$\vartheta(Z) = \sum_{g \in \mathbb{Z}^n} \exp \pi i Z[g]$$

defines a holomorphic function on $\mathbb{H}_n$.

3.5 Proposition. *There exists a multiplier system*

$$v_\vartheta : \Gamma_{n,\vartheta} \longrightarrow \mathbb{C}^\bullet$$

of weight 1/2, such that the transformation formula

$$\vartheta(M\langle Z\rangle) = v_\vartheta(M)\sqrt{\det(CZ+D)}\vartheta(Z)$$

holds for all $M \in \Gamma_{n,\vartheta}$. One has

$$v_\vartheta(M)^8 = 1.$$

It is possible to express $v_\vartheta(M)$ by Gauss sums. We shall do this in sec.5 for M with invertible D.

Let

$$v : \Gamma \longrightarrow \mathbb{C}^\bullet$$

now be an arbitrary multiplier system of weight $r/2$. Then the function

$$\Gamma \cap \Gamma_{n,\vartheta} \longrightarrow \mathbb{C}^\bullet,$$
$$M \longmapsto v(M)/v_\vartheta(M)^r,$$

is obviously a character. The result 1.4 of MENNICKE implies: In case $n > 1$ there exists a congruence subgroup $\Gamma_0 \subset \Gamma \cap \Gamma_{n,\vartheta}$ such that

$$v|\Gamma_0 = v_\vartheta^r|\Gamma_0.$$

We make this as an assumption (even in the case $n = 1$).

3.6 Assumption. *All multiplier systems which we consider agree on a suitable congruence subgroup with a power of the theta multiplier system.*

3.7 Definition. *A real matrix M is called projectively rational if there exists a number $t \neq 0$ such that tM is rational (i.e. has rational entries).*

3.8 Remark. *Assume that $\Gamma \subset \mathrm{Sp}(n, \mathbb{R})$ is a congruence subgroup and $N \in \mathrm{Sp}(n, \mathbb{R})$ is projectively rational. Then*

$$\Gamma^N = N^{-1}\Gamma N$$

is also a congruence subgroup.

Using 3.4 one easily proves:

3.9 Lemma. *Let*

$$\Gamma \subset \mathrm{Sp}(n, \mathbb{R})$$

be a congruence subgroup and

$$v : \Gamma \longrightarrow \mathbb{C}^{\bullet}$$

be a multiplier system. Then

$$v^N : \Gamma^N \longrightarrow \mathbb{C},$$

$$v^N(N^{-1}MN) = v(M)w(M,N)w(N,N^{-1}MN), \quad M \in \Gamma,$$

is again a multiplier system of the same weight.

We call v^N the conjugate multiplier system. Of course one has to show that v^N satisfies again our assumption 3.6. This has to be done only for the theta multiplier system v_ϑ itself.

3.10 Lemma. *Each conjugate of the theta multiplier system satisfies assumption 3.6.*

Corollary. *The class of multiplier systems which satisfy assumption 3.6 is stable under conjugation.*

This result follow froms from MENNICKE's theorem but also from the general theta transformation formalism (s. sec.6). This formalism also implies the following refined result [Ig1].

3.11 Lemma. *The theta multiplier system and all its conjugates*

$$v_\vartheta^M, \quad M \in \Gamma_n,$$

agree on the group $\Gamma_n[4,8]$. v_ϑ^2 is trivial on this group.

4   Siegel modular forms

We introduce the notion of a vector valued Siegel modular form with possibly half integral weight.

Throughout the whole notes we fix a rational representation

$$\varrho_0 : \mathrm{GL}(n, \mathbb{C}) \longrightarrow \mathrm{GL}(\mathcal{Z}), \quad \dim_{\mathbb{C}} \mathcal{Z} < \infty.$$

"Rational" means that there exists an integer $k \in \mathbb{Z}$ such that

$$A \longmapsto (\det A)^{-k}\varrho_0(A)$$

is polynomial. There always exists a biggest number k with this property. We call this number

$$k = k(\varrho_0)$$

the **weight** of ϱ_0. (We shall see later that k is nothing but the last component of the "highest weight" of ϱ_0.)

4.1 Definition. *A representation*

$$\varrho_0 : \mathrm{GL}(n, \mathbb{C}) \longrightarrow \mathrm{GL}(\mathcal{Z})$$

*is called **reduced** if its weight is 0,*
equivalently:
ϱ_0 is polynomial and does not vanish on the determinant surface "$\det A = 0$".

We also fix an integer

$$r \in \mathbb{Z}.$$

We will be interested in pairs

$$(\varrho_0, r).$$

4.2 Definition.   *Two pairs are called equivalent,*

$$(\varrho_0, r) \equiv (\varrho_0', r'),$$

if and only if

   a) $r \equiv r' \bmod 2$,

   b) $\varrho_0(A) \cdot \det A^{\frac{r-r'}{2}} = \varrho_0'(A)$.
The equivalence class is denoted by

$$\varrho = [\varrho_0, r].$$

Each pair contains precisely one representative (ϱ_0, r), such that ϱ_0 is reduced. We call r the weight of ϱ.

   If r is even, $\varrho = [\varrho_0, r]$ is determined uniquely by the representation

$$\varrho(A) = \varrho_0(A) \det A^{r/2}.$$

The equivalence class should be considered only as a compensation for this if r is odd. Hence we use the symbolic notation

$$\text{"}\varrho(A) = \varrho_0(A) \det A^{r/2}\text{"}$$

in all cases. In any case, if the square root of $\det(CZ + D)$ has been chosen, the expression

$$\varrho(CZ + D) := \varrho_0(CZ + D)\sqrt{\det(CZ + D)}^{r}$$

is well defined and depends only on ϱ.

   Let Γ be a congruence subgroup of $\mathrm{Sp}(n, \mathbb{R})$ and

$$v : \Gamma \longrightarrow \mathbb{C}^{\bullet}$$

a multiplier system of weight $r/2$. Then

$$v(M)\varrho(CZ + D)$$

satisfies the cocycle relation. Hence it makes sense to consider holomorphic functions

$$f : \mathbb{H}_n \longrightarrow \mathcal{Z}$$

with values in the representation space $\mathcal{Z}$, which have the transformation property

$$f(M\langle Z\rangle) = v(M)\varrho(CZ + D)f(Z)$$

for all $M \in \Gamma$. It follows from our assumption on the multiplier system v that f is periodic with respect to some rational lattice L of symmetric real matrices and hence admits a Fourier expansion

$$f(Z) = \sum_{T \in L^*} \exp 2\pi i\sigma(TZ).$$

Here

$$L^* = \{T = T' \text{ real}; \quad \sigma(TX) \in \mathbb{Z} \text{ for all } X \in L\}$$

denotes the dual lattice of L. Besides f, we have to consider the conjugate functions

$$f^N(Z) := (f|_\varrho N)(Z) = \varrho(CZ + D)^{-1}f(N\langle Z\rangle),$$

where $N \in \mathrm{Sp}(n, \mathbb{R})$ is some projective rational symplectic matrix. From 3.4 follows that f^N inherits from f a transformation formula for the conjugate group $N^{-1}\Gamma N$. More precisely:

$$f^N(M\langle Z\rangle) = v^N(M)\varrho(CZ + D)f^N(Z)$$

for all $M \in \Gamma^N$, where v^N denotes the conjugate multiplier system (3.9).

It follows that f^N admits a Fourier expansion

$$f^N(Z) = \sum_T a^N(T) \exp 2\pi i\sigma(TZ),$$

where T runs through some rational lattice depending on f and on N.

4.3 Definition. *Let $\Gamma \in \mathrm{Sp}(n, \mathbb{R})$ be a congruence subgroup and v a multiplier system of weight $r/2$ on Γ. A holomorphic function*

$$f : \mathbb{H}_n \longrightarrow \mathcal{Z}$$

is called a modular form with respect to Γ, ϱ and v, if the transformation formula

$$f(M\langle Z\rangle) = v(M)\varrho(CZ + D)f(Z) \quad \text{for } M \in \Gamma$$

is satisfied and if

$$a^N(T) \neq 0 \Longrightarrow T \geq 0$$

for all projective rational N.

Half the weight of ϱ is called the weight of f.

We denote by

$$\Omega_n \subset \operatorname{Sp}(n, \mathbb{R})$$

the group of all projectively rational matrices and by

$$\Omega_{n,0} \subset \operatorname{Sp}(n, \mathbb{R})$$

the subgroup consisting of all elements

$$M = \begin{pmatrix} A & B \\ 0 & D \end{pmatrix} \in \Omega_n.$$

4.4 Remark. *The last condition in Definition 4.3 depends only on the double coset*

$$\Gamma \backslash \Omega_n / \Omega_{n,0}.$$

An easy lemma states ([Fr4], II.6.2):

4.5 Lemma. *Each coset of*

$$\Omega_n / \Omega_{n,0}$$

contains an integral element $M \in \operatorname{Sp}(n, \mathbb{Z})$. *In particular, the set*

$$\Gamma \backslash \Omega_n / \Omega_{n,0}$$

is finite.

Corollary. *In the case of the full modular group it is sufficient to consider*

$$N = E$$

in 4.3.

An important observation of KOECHER (which we don't need) states:

In the case $n > 1$ the last condition in 4.3 follows from the transformation property. Hence it can be omitted in 4.3.

4.6 Definition. *A modular form is called a cusp form if*

$$a^N(T) \neq 0 \Longrightarrow T > 0.$$

Again it is sufficient to consider a system of representatives

$$N : \quad \Gamma \backslash \Omega_n / \Omega_{n,0}.$$

Notation

$[\Gamma, \varrho, v]$ = space of all modular forms in sense of 4.3,

$[\Gamma, \varrho, v]_0$ = subspace of cusp forms.

It is well-known that these spaces are finite dimensional. We need a certain vanishing theorem for these spaces.

4.7 Proposition. *1)   Each modular form of negative weight vanishes.*

2)   Each modular form of weight 0 is constant.

An elementary proof of this result can be found in [Fr6]. A much more general vanishing theorem has been proved by Weissauer [We1] using non-trivial representation theoretic methods.

5   The theta multiplier system

In these notes we want to include modular forms of half integral weight. For this purpose we need some information about the theta multiplier system. Its theory goes back to the 19th century. We refer to the classical report of KRAZER and WIRTINGER [KW]. Part of this theory has been reorganized by IGUSA [Ig1-3]. It is nevertheless difficult to bring together from the literature all the material which we need in these notes. Therefore we have included complete proofs for all that we need.

The easiest theta series of degree n is

$$\vartheta(Z) := \sum_{g \in \mathbb{Z}^n} \exp \pi i Z[g].$$

It is trivial that this series converges uniformly on domains

$$Y - \delta E \geq 0,\ \delta > 0.$$

Hence it defines a holomorphic function on the half space $\mathbb{H}_n$.

5.1 Proposition. *The theta series*

$$\vartheta(Z) := \sum_{g \in \mathbb{Z}^n} \exp \pi i Z[g]$$

is a modular form of weight $1/2$ with respect to the theta group

$$\Gamma_{n,\vartheta} = \{M = \begin{pmatrix} A & B \\ C & D \end{pmatrix} \in \mathrm{Sp}(n, \mathbb{Z});\ AB'\ and\ CD'\ have\ even\ diagonal\ entries\}$$

and with respect to a certain multiplier system v_ϑ — the so-called theta multiplier system,

$$\vartheta \in [\Gamma_{n,\vartheta}, 1/2, v_\vartheta],$$

i.e.

$$\vartheta\big((AZ + B)(CZ + D)^{-1}\big) = v_\vartheta(M)\sqrt{\det(CZ + D)}\vartheta(Z)$$

for all $M \in \Gamma_{n,\vartheta}$.

For the proof one considers more general theta series

$$\vartheta[\mathbf{m}](Z) = \sum_{g \in \mathbb{Z}^n} \exp \pi i \{Z[g + a] + 2(g + a)'b\}.$$

The characteristic

$$\mathbf{m} = \begin{pmatrix} a \\ b \end{pmatrix};\quad a \in \mathbb{C}^n,\ b \in \mathbb{C}^n$$

is a column of $2n$ complex numbers. It is easy to show that $\vartheta[\mathbf{m}](Z)$ defines a holomorphic function on

$$\mathbb{H}_n \times \mathbb{C}^{2n}.$$

5.2 Remark. *Up to a constant factor, $\vartheta\,[\mathbf{m}]$ depends only on $\mathbf{m} \bmod 1$; more precisely*

$$\vartheta[\mathbf{m}] = \exp\big(2\pi i a'(b - \widetilde{b})\big)\,\vartheta[\widetilde{\mathbf{m}}]\ if\ \mathbf{m} \equiv \widetilde{\mathbf{m}}\ \bmod 1.$$

We define an affine action of the Siegel modular group on the set of characteristics.

5.3 Definition. *For*

$$M \in \Gamma_n \text{ and } \mathbf{m} \in \mathbb{C}^{2n}$$

we define

$$M\{\mathbf{m}\} = (M')^{-1} \cdot \mathbf{m} + \frac{1}{2} \begin{pmatrix} (CD')_0 \\ (AB')_0 \end{pmatrix}.$$

In this context we denote by

$$S_0 = \begin{pmatrix} s_{11} \\ \vdots \\ s_{nn} \end{pmatrix}$$

the column built out of the diagonal elements of a symmetric matrix S.

5.4 Remark. *One has*

a) $E\{\mathbf{m}\} = \mathbf{m}$,
b) $(MN)\{\mathbf{m}\} \equiv M\{N\{\mathbf{m}\}\} \mod 1$,

i.e Γ_n *acts on the set*

$$(\mathbb{C}/\mathbb{Z})^{2n}.$$

Proof. With the notations

$$M = \begin{pmatrix} A_1 & B_1 \\ C_1 & D_1 \end{pmatrix}; \quad N = \begin{pmatrix} A_2 & B_2 \\ C_2 & D_2 \end{pmatrix}$$

one has to show:

$$\begin{pmatrix} D_1 & -C_1 \\ -B_1 & A_1 \end{pmatrix} \cdot \begin{pmatrix} (C_2 D_2')_0 \\ (A_2 B_2')_0 \end{pmatrix} + \begin{pmatrix} (C_1 D_1')_0 \\ (A_1 B_1')_0 \end{pmatrix}$$
$$\equiv \begin{pmatrix} (C_1 A_2 + D_1 C_2)(C_2 B_2 + D_1 D_2)')_0 \\ ((A_1 A_2 + B_1 C_2)(A_1 B_2 + B_1 D_2)')_0 \end{pmatrix} \mod 2.$$

This follows easily from the symplectic relations combined with the fact that for symmetric integral matrices S and for integral vectors g the congruence relation

$$S[g] \equiv S_0' g \mod 2$$

holds.

5.5 Proposition. *We have*

$$\vartheta[M\{\mathbf{m}\}](M\langle Z\rangle) = v(M, \mathbf{m})\sqrt{\det(CZ + D)}\vartheta[\mathbf{m}](Z)$$

for all $M \in \Gamma_n = \mathrm{Sp}(n, \mathbb{Z})$. *Here* $v(M, \mathbf{m})$ *denotes a system of complex numbers which is independent of* Z.
One furthermore has

   $\mathbf{m}$ *real* $\implies |v(M, \mathbf{m})| = 1$,
   $\mathbf{m}$ *integral* $\implies v(M, \mathbf{m})^8 = 1$.

Because of the remarks 5.2 and 5.4 this proposition has to be proved only for the generators of the group Γ_n, hence for

$$M = \begin{pmatrix} E & S \\ 0 & E \end{pmatrix}; \quad \begin{pmatrix} 0 & E \\ -E & 0 \end{pmatrix}.$$

1) $$M = \begin{pmatrix} E & S \\ 0 & E \end{pmatrix}.$$

The formulae

$$S[g+a] = S[g] + 2a'Sg + S[a]$$

and

$$S[g] \equiv S_0'g \quad \mod 2$$

show

$$\vartheta[\mathbf{m}](Z+S) = \exp(\pi i S[a])\vartheta[\tilde{\mathbf{m}}](Z)$$

where

$$\tilde{\mathbf{m}} = \begin{pmatrix} a \\ b + Sa + \frac{1}{2}S_0 \end{pmatrix}.$$

2)   The theta inversion formula states

$$\vartheta\begin{bmatrix} a \\ b \end{bmatrix}(-Z^{-1}) = \exp(2\pi i a'b)\sqrt{\det(Z/i)}\,\vartheta\begin{bmatrix} -b \\ a \end{bmatrix}(Z),$$

where the branch of the square root is defined by

$$\sqrt{\det(Z/i)} = +1 \text{ if } Z = iE.$$

We will prove this inversion formula a little later in a more general context (chap.II, sec.2).

It is natural to ask for the $\mathbf{m}$–dependency of $v(M, \mathbf{m})$.

5.6 Proposition. *We have*

$$v(M, \mathbf{m}) = v(M)\exp\big(2\pi i \Phi_{\mathbf{m}}(M)\big),$$

where

$$-2\Phi_{\mathbf{m}}(M) = (B'D)[a] + (A'C)[b] - 2a'B'Cb - (Da - Cb)'(AB')_0.$$

Proof.

First step: Reduction to $a = 0$.

One uses the trivial formula

$$\vartheta\begin{bmatrix} a \\ b \end{bmatrix}(Z) = \exp\big(\pi i(Z[a] + 2a'b)\big)\vartheta\begin{bmatrix} 0 \\ b + Za \end{bmatrix}(Z)$$

and obtains

$$\vartheta[M\{\mathbf{m}\}](M\langle Z\rangle) = v(M,\mathbf{m})\sqrt{\det(CZ+D)}\cdot$$

$$\exp\big(\pi i(Z[a] + 2a'b)\big)\vartheta\begin{bmatrix} 0 \\ b + Za \end{bmatrix}(Z)$$

$$= v(M,\mathbf{m})v\Big(M, \begin{matrix} 0 \\ b + Za \end{matrix}\Big)^{-1}\exp\big(\pi i(Z[a] + 2a'b)\cdot$$

$$\vartheta\Big[M\Big\{\begin{matrix} 0 \\ b + Za \end{matrix}\Big\}\Big](M\langle Z\rangle).$$

This formula leads us to compare

$$\vartheta[M\{\mathbf{m}\}](W) \text{ and } \vartheta\Big[M\Big\{\begin{matrix} 0 \\ b + Za \end{matrix}\Big\}\Big](W),$$

where $W = M\langle Z\rangle$.

a)
$$\vartheta[M\{\mathbf{m}\}](W) = \vartheta\begin{bmatrix} Da - Cb + \frac{1}{2}(CD')_0 \\ -Ba + Ab + \frac{1}{2}(AB')_0 \end{bmatrix}(W)$$

$$= \exp\bigg(\pi i\Big(W\big[Da - Cb + \frac{1}{2}(CD')_0\big] +$$

$$2\big(Da - Cb + \frac{1}{2}(CD')_0\big)'\big(-Ba + Ab + \frac{1}{2}(AB')_0\big)\Big)\bigg)\cdot$$

$$\vartheta\begin{bmatrix} 0 \\ -Ba + Ab + \frac{1}{2}(AB')_0 + W\big(Da - Cb + \frac{1}{2}(CD')_0\big) \end{bmatrix}(W).$$

b)
$$\vartheta\Big[M\Big\{\begin{matrix} 0 \\ b + Za \end{matrix}\Big\}\Big](W) = \vartheta\begin{bmatrix} -C(b + Za) + \frac{1}{2}(CD')_0 \\ A(b + Za) + \frac{1}{2}(AB')_0 \end{bmatrix}(W) =$$

$$\exp \pi i\bigg(W\big[-C(b + Za) + \frac{1}{2}(CD')_0\big] +$$

$$2\big(-C(b + Za) + \frac{1}{2}(CD')_0\big)'\big(A(b + Za) + \frac{1}{2}(AB')_0\big)\bigg)\cdot$$

$$\vartheta\begin{bmatrix} 0 \\ A(b + Za) + \frac{1}{2}(AB')_0 + W\big(-C(b + Za) + \frac{1}{2}(CD')_0\big) \end{bmatrix}(W).$$

But the two remaining theta functions agree, because

$$-Ba + Ab + \frac{1}{2}(AB')_0 + W\big(Da - Cb + \frac{1}{2}(CD')_0\big) -$$

$$A(b + Za) - \frac{1}{2}(AB')_0 - W\big(-C(b + Za) + \frac{1}{2}(CD')_0\big)$$

$$= -Ba + WDa + WCZa - AZa = -(Ba + AZa) + W(D + CZ)a = 0.$$

Now we obtain

$$\exp\Big(\pi i\Big(W\big[Da - Cb + \tfrac{1}{2}(CD')_0\big]+$$

$$2\big(Da - Cb + \tfrac{1}{2}(CD')_0\big)'\big(-Ba + Ab + \tfrac{1}{2}(AB')_0\big)\Big)\Big)$$

$$= v(M,\mathbf{m})v\left(M,\ {}^{\ \ 0}_{b+Za}\right)^{-1}$$

$$\exp \pi i(Z[a] + 2a'b)$$

$$\exp\Big(\pi i\Big(W\big[-C(b + Za) + \tfrac{1}{2}(CD')_0\big]+$$

$$2\big(-C(b + Za) + \tfrac{1}{2}(CD')_0\big)'\big(A(b + Za) + \tfrac{1}{2}(AB')_0\big)\Big).$$

This formula expresses $v(M,\mathbf{m})$ as product of $v\left(M,\ {}^{\ \ 0}_{b+Za}\right)$ and some elementary factors. A careful but trivial calculation, which can be left to the reader, now gives the reduction of proposition 5.6 to the case $a = 0$ and $b + Za$ instead of b.

Second step: Proof of proposition 5.6 in the case a=0.

We have to prove that the function

$$\frac{\vartheta[\mathbf{m}](Z)\exp\big(2\pi i\Phi_\mathbf{m}(M)\big)}{\vartheta[M\{\mathbf{m}\}](M\langle Z\rangle)};\quad \mathbf{m} = \binom{0}{b}$$

is independent of b (for fixed Z and M). We write w instead of b and obtain

$$f(w) := \vartheta[\mathbf{m}](Z) = \sum_{g\in\mathbb{Z}^n} \exp \pi i\big(Z[g] + 2w'g\big).$$

This function obviously satisfies the functional equations

a) $f(w + h) = f(w),$

b) $f(w + Zh) = \exp\big(-\pi i(Z[h] + 2w'h)\big)f(w)$ for $h \in \mathbb{Z}^n.$

We notice that the entire function f is characterized by those two functional equations up to a constant factor (which may depend on Z):
From equation a) one obtains a Fourier expansion

$$f(w) = \sum_{h\in\mathbb{Z}^n} a_h \exp 2\pi i h'w.$$

Equation b) shows

$$a_h = a_0 \exp \pi i Z[h].$$

For the second step it is sufficient to prove equations a), b) for the function

$$g(w) := \frac{\vartheta[M\{\mathbf{m}\}](M\langle Z\rangle)}{\exp 2\pi i\Phi_\mathbf{m}(M)}$$

instead of w. This is a not quite trivial calculation using the symplectic relations, and is left to the reader.

The transformation formula can be simplified considerably if M is contained in the theta group.

5.7 Theorem. *The modified theta series*

$$\widetilde{\vartheta}[m] = e^{-\pi i a' b}\vartheta[m]$$

satisfies for all $M \in \Gamma_{n,\vartheta}$ *the transformation formula*

$$\widetilde{\vartheta}\left[\begin{pmatrix} D & -C \\ -B & A \end{pmatrix} \cdot m\right](M\langle Z\rangle) = v_\vartheta(M)\det(CZ+D)^{1/2}\widetilde{\vartheta}[m](Z).$$

The number $v_\vartheta(M)$ *is a eighth root of unity. It is of course independent of* **m** *(but depends on the choice of the square root of* $\det(CZ+D)$*).*

Warning. If M is contained in the theta group, then

$$M\{m\} \equiv M'^{-1}m \pmod 1.$$

Both sides are not equal. Hence $\vartheta[M\{m\}]$ and $\vartheta[M'^{-1}m]$ may be different. An easy calculation shows

$$\vartheta[M\{m\}] = \vartheta[M'^{-1}m] \cdot \exp\pi i(Da - Cb)'(AB')_0.$$

From this formula together with 5.6 follows 5.7.

Appendix to sec.5
The theta multiplier system as Gauss sum

We want to express the theta multiplier system

$$v_\vartheta(M); \quad M = \begin{pmatrix} A & B \\ C & D \end{pmatrix} \in \Gamma_{n,\vartheta}$$

for invertible D as a Gauss sum. First we fix a choice of the square root of $\det(CZ+D)$ in this case. We denote by

$$h(Z) = \sqrt{\det(Z/i)}$$

the unique holomorphic function on $\mathbb{H}_n$ with the properties

a) $h(Z)^2 = \det(Z/i)$,

b) $h(iE) = +1$

and define

$$\sqrt{\det(CZ+D)} := \sqrt{\det D}\, h(Z)h(-Z^{-1} - D^{-1}C),$$

where the square root of $\det D$ is taken as

$$\sqrt{\det D} > 0 \quad \text{if } \det D > 0,$$
$$-i\sqrt{\det D} > 0 \quad \text{if } \det D < 0.$$

Using this convention, we have

5.8 Proposition. *Assume*

$$M = \begin{pmatrix} A & B \\ C & D \end{pmatrix} \in \Gamma_{n,\vartheta}; \quad \det D \neq 0.$$

Then

$$v_\vartheta(M) = (\det D)^{-1/2} \sum_{h \in \mathbb{Z}^n / D\mathbb{Z}^n} \exp \pi i (BD^{-1})[h].$$

Proof. We start with the defining formula

$$\vartheta(M\langle Z\rangle) = v_\vartheta(M) \det(CZ + D)^{1/2} \vartheta(Z).$$

We multiply both sides by $\det(Z/i)^{1/2}$ and specialize Z to 0. More precisely, Z will run through the sequence

$$\frac{1}{m} iE; \quad m = 1, 2, \ldots.$$

From the theta inversion formula

$$\lim_{Z \to 0} \det(Z/i)^{1/2} \vartheta(Z) = \lim_{Z \to 0} \vartheta(-Z^{-1}) = 1.$$

This gives v_ϑ as a value at a "boundary point", namely

$$v_\vartheta(M) = \det(D)^{-1/2} \lim_{Z \to 0} \vartheta(M\langle Z\rangle) \cdot \det(Z/i)^{1/2}.$$

The computation of this boundary value is based on the simple formula

$$\begin{pmatrix} A & B \\ C & D \end{pmatrix} = \begin{pmatrix} E & R \\ 0 & E \end{pmatrix} \begin{pmatrix} D'^{-1} & 0 \\ C & D \end{pmatrix}; \quad R = BD^{-1},$$

which follows immediately from the symplectic relations 1.1. It can be considered as a special case of the so-called Bruhat decomposition. It implies

$$M\langle Z\rangle = W + R; \quad W = D'^{-1} Z (CZ + D)^{-1}.$$

This gives us

$$\vartheta(M\langle Z\rangle) = \vartheta(W + R)$$

$$= \sum_{h \in \mathbb{Z}^n / D\mathbb{Z}^n} \exp \pi i R[h] \sum_{g \equiv h \bmod D} \exp \pi i W[g]$$

$$= \sum_{h \in \mathbb{Z}^n / D\mathbb{Z}^n} \exp \pi i R[h] \vartheta\begin{bmatrix} D^{-1}h \\ 0 \end{bmatrix} (W[D]).$$

Applying the inversion formula again, we obtain

$$\vartheta\begin{bmatrix} D^{-1}h \\ 0 \end{bmatrix}(W[D]) = \det(W[D]/i)^{-1/2} \vartheta\begin{bmatrix} 0 \\ D^{-1}h \end{bmatrix}(-W[D]^{-1}).$$

Using

$$-W[D]^{-1} = -Z^{-1} - D^{-1}C$$

we obtain

$$\lim_{Z \to 0} \vartheta\begin{bmatrix} 0 \\ D^{-1}h \end{bmatrix}(-W[D]^{-1}) = 1.$$

Now the claimed formula follows immediately.

The symplectic Gauss sum as generalized Legendre symbol

5.9 Definition. *Let C,D be two integral $n \times n$-matrices such that $\det D$ is not equal to 0 and such that CD' is even. We define*

$$G_D(C) := \sum_{g \in \mathbb{Z}^n / D\mathbb{Z}^n} \exp \pi i g' C D^{-1} g.$$

Each term in the sum remains unchanged if one replaces g by $g + Dh$, $h \in \mathbb{Z}^n$. The sum is therefore well-defined and contains $|\det D|$ elements. We are mainly interested in the case where (C, D) is the second row of a matrix of the theta group and the sum in this case will be called a **symplectic Gauss sum**. This sum is very classical in the case $n = 1$, C even

$$G(c,d) = \sum_{t=0}^{d-1} e^{\frac{2\pi i c}{d} t^2} \quad (= G_d(2c)).$$

It occurs in the theory of the quadratic reciprocity law and can be expressed by means of the (generalized) Legendre symbol, whose definition we recall.

5.10 Definition (of the Legendre symbol). *Assume that a is an integer and that p is an odd prime, which does not divide a. Then*

$$\left(\frac{a}{p}\right) := \left\{ \begin{array}{ll} 1 & \textit{if } a \textit{ is a square} \bmod p; \\ -1 & \textit{otherwise.} \end{array} \right.$$

5.11 Definition (of the Jacobi symbol). *Assume that a,b are two coprime integers, different from 0. Assume furthermore that b is odd. Let*

$$b = \epsilon p_1 \cdots p_m; \quad \epsilon \in \{1, -1\},$$

be the decomposition of b into primes. One defines

$$\left(\frac{a}{b}\right) := \prod_{\nu}^{m} \left(\frac{a}{p_\nu}\right).$$

Obviously

$$\left(\frac{a}{b}\right) = \left(\frac{a}{|b|}\right).$$

5.12 Lemma. *Asume that a,b are coprime integers different from 0 and that b is odd. Then*

$$G(a,b) = \left(\frac{a}{b}\right) G(1,b),$$

and

$$G(1,b) = \left\{ \begin{array}{ll} |b|^{1/2} & \textit{if } b \equiv 1 \mod 4, \\ i|b|^{1/2} & \textit{if } b \equiv -1 \mod 4. \end{array} \right.$$

For a proof we refer to [Ha], [Ei]. We end this section with the main properties of the Jacobi symbol, especially the quadratic reciprocity law.

5.13 Proposition. *The Jacobi symbol has the following properties:*

1)
$$\left(\frac{-1}{b}\right) = (-1)^{\frac{b-1}{2} + \frac{\text{sgn}\, b - 1}{2}},$$

2)
$$\left(\frac{2}{b}\right) = (-1)^{(b^*-1)/4}, \quad b^* = (-1)^{\frac{b-1}{2}} \cdot b,$$

3)
$$\left(\frac{a}{b}\right)\left(\frac{b}{a}\right) = (-1)^{\frac{a-1}{2}\frac{b-1}{2} + \frac{\text{sgn}\, a - 1}{2}\frac{\text{sgn}\, b - 1}{2}} \quad \text{if } a \text{ is odd.}$$

Sometimes a further generalization of the Legendre symbol is useful:

5.14 Remark (Definition of the Kronecker symbol). *There exists a unique extension*
$$\left(\frac{a}{b}\right)$$
of the Jacobi symbol including even b as long as $a \equiv 1 \bmod 4$ (a, b are coprime integers both different from 0), such that the reprocity law 3) in 5.13 is true also for even b (and $a \equiv 1 \bmod 4$).

The symplectic Gauss sum should be considered as a generalization of the Jacobi symbol. This is justified because it satisfies a reprocity law [Sta]. One would like to express it by means of the usual Jacobi symbol. In some sense this can be done but not in an easy way. For details we refer to the papers [Sta], [Sty1], [Sty2]. Special cases will be done later (s. chap.II sec.7).

6   The Siegel Φ-operator

The Siegel Φ-operator reduces the degree of a modular form. It is often very useful for induction arguments. In the vector valued case, a good treatment of this operator needs the theory of highest weights of rational representations of $GL(n, \mathbb{C})$. In an appendix to this section we describe this theory without proofs. This generalization of the Φ-operator to the vector valued case is due to WEISSAUER [We1].

Let
$$f : \mathbb{H}_n \longrightarrow \mathcal{Z} \quad (\dim_{\mathbb{C}} \mathcal{Z} < \infty)$$
by a function on the half space of degree n. The Φ-operator is defined by
$$f|\Phi : \mathbb{H}_{n-1} \longrightarrow \mathcal{Z},$$
$$(f|\Phi)(Z) = \lim_{t \to \infty} f\begin{pmatrix} Z & 0 \\ 0 & it \end{pmatrix},$$
if this limit exists.

6.1 Remark. *Let $\mathcal{L}$ be a rational lattice of symmetric $n \times n$-matrices and f a Fourier series of the type*

$$f(Z) = \sum_{T \in \mathcal{L}^\bullet} a(T) \exp 2\pi i\sigma(TZ),$$

$$a(T) \neq 0 \Longrightarrow T \geq 0,$$

which converges on $\mathbb{H}_n$. Then the Φ-operator can be applied term by term. One obtains

$$(f|\Phi)(Z) = \lim_{t \to \infty} f \begin{pmatrix} Z & 0 \\ 0 & it \end{pmatrix}$$

$$= \sum_{\begin{pmatrix} T & 0 \\ 0 & 0 \end{pmatrix} \in \mathcal{L}^\bullet} a \begin{pmatrix} T & 0 \\ 0 & 0 \end{pmatrix} \exp 2\pi i\sigma(TZ).$$

We denote by Φ^j $(0 \leq j \leq n)$ the j-fold application of the Φ-operator. By means of the Fourier expansion it is easy to show (under the assumption of 6.1) that

$$\left(f|\Phi^j\right)(Z) = \lim_{t \to \infty} \begin{pmatrix} Z & 0 \\ 0 & itE \end{pmatrix} \qquad (Z \in \mathbb{H}_{n-j}).$$

The Φ-operator especially can be applied to Siegel modular forms

$$f \in [\Gamma, \varrho, v].$$

The function

$$g := f|\Phi^j$$

inherits from f a functional equation:

$$g(M\langle Z \rangle) = v(\widetilde{M})\varrho \begin{pmatrix} CZ + D & 0 \\ 0 & E \end{pmatrix} g(Z)$$

for all

$$M = \begin{pmatrix} A & B \\ C & D \end{pmatrix} \in \mathrm{Sp}(n-j, \mathbb{R})$$

such that

$$\widetilde{M} = \begin{pmatrix} A & 0 & B & 0 \\ 0 & E & 0 & 0 \\ C & 0 & D & 0 \\ 0 & 0 & 0 & E \end{pmatrix} \in \Gamma.$$

The representation

$$A \longmapsto \varrho_0 \begin{pmatrix} A & 0 \\ 0 & E \end{pmatrix}$$

is no longer irreducible. But the values of g are contained in some distinguished subspace $\mathcal{Z}_0 \subset \mathcal{Z}$:

We denote by $N_{n,j}$ the abelian group

$$N_{n,j} = \left\{ \begin{pmatrix} E^{(n-j)} & B \\ 0 & E^{(j)} \end{pmatrix}; \quad B \in \mathbb{C}^{(n-j,j)} \right\}$$

and by

$$\mathcal{Z}_0 = \mathcal{Z}^{N_{n,j}}$$

the invariant subspace consisting of all vectors $z \in \mathcal{Z}$, such that

$$\varrho_0(A)z = z \quad \text{for all } A \in N_{n,j}.$$

6.2 Lemma. *Let $f \in [\Gamma, \varrho, v]$ be a modular form and $0 \le j \le n$. The values of $f|\Phi^j$ are contained in the subspace*

$$\mathcal{Z}_0 = \mathcal{Z}^{N_{n,j}}.$$

They generate a subspace of $\mathcal{Z}_0$ which is invariant under the substitutions

$$\varrho_0 \begin{pmatrix} E^{(n-j)} & 0 \\ 0 & D \end{pmatrix}; \quad D \in \mathrm{SL}(j, \mathbb{C}).$$

Proof. The equations

$$f(Z[U]) = \varrho(U^{-1})f(Z),$$
$$a(T[U]) = \varrho(U')a(T)$$

and

$$\begin{pmatrix} T & 0 \\ 0 & 0 \end{pmatrix} \begin{bmatrix} E & 0 \\ B' & D' \end{bmatrix} = \begin{pmatrix} T & 0 \\ 0 & 0 \end{pmatrix}$$

show the claimed invariance properties for all B and D under certain congruence conditions. But the identities are polynomial and hence are valid for all B and all $D \in \mathrm{SL}(j, \mathbb{C})$.

Now we make use of the theory of highest weights, which is explained in the appendix to this section without proofs.

Let

$$(r_1, \ldots, r_n)$$

be the highest weight of ϱ_0. In the appendix to this section we will see (6.11):

6.3 Proposition. *Assume that*

$$g = f|\Phi^j \qquad (f \in [\Gamma, \varrho, v])$$

does not vanish identically. Then the representation

$$\varrho_0|\Phi^j : \mathrm{GL}(n-j, \mathbb{C}) \longrightarrow \mathrm{GL}(\mathcal{Z}_0),$$

$$A \longmapsto \varrho_0 \begin{pmatrix} A & 0 \\ 0 & E \end{pmatrix} \Big| \mathcal{Z}_0$$

is irreducible and has the highest weight

$$(r_1, \ldots, r_{n-j}).$$

Moreover one has

$$r_{n-j+1} = \ldots = r_n.$$

We introduce some further notations:

1) $(\varrho|\Phi^j)(A) = (\varrho_0|\Phi^j)(A)(\det A)^{r/2}.$

2)   We consider the imbedding

$$\mathrm{Sp}(n-j,\mathbb{R}) \longrightarrow \mathrm{Sp}(n,\mathbb{R}),$$

$$M = \begin{pmatrix} A & B \\ C & D \end{pmatrix} \longmapsto \widetilde{M} = \begin{pmatrix} A & 0 & B & 0 \\ 0 & E & 0 & 0 \\ C & 0 & D & 0 \\ 0 & 0 & 0 & E \end{pmatrix}.$$

We denote by

$$\Gamma|\Phi^j = \{M; \quad \widetilde{M} \in \Gamma\}$$

the inverse image of Γ. Of course, $\Gamma|\Phi^j$ is again a congruence subgroup. Finally we define the multiplier system $v|\Phi^j$ on $\Gamma|\Phi^j$ by

$$(v|\Phi^j)(M)\sqrt{\det(CZ+D)}^r = v(\widetilde{M})\sqrt{\det\left(\widetilde{C}\begin{pmatrix} Z & 0 \\ 0 & iE \end{pmatrix} + \widetilde{D}\right)}^r$$

(The two square roots occuring in this equation agree after a suitable choice of their sign).

We collect:

6.4 Proposition. *Assume that $f \in [\Gamma, \varrho, v]$ is a modular form with the property*

$$f|\Phi^j \neq 0.$$

Then the highest weight $(r_1, \ldots, r_n)$ of ϱ has the property $r_{n-j+1} = \ldots = r_n$. The Φ-operator defines a map

$$[\Gamma, \varrho, v] \longrightarrow [\Gamma|\Phi^j, \varrho|\Phi^j, v|\Phi^j],$$

where $\varrho_0|\Phi^j$ is an irreducible representation with highest weight $(r_1, \ldots, r_{n-j})$.

Appendix to Sec.6
Highest weights

We recall without proofs some basic facts about rational representations of the group

$$G(n) = \mathrm{GL}(n, \mathbb{C}),$$

more generally of the group

$$G = G(n_1) \times \ldots \times G(n_k).$$

Proofs can be found in [NS].

A rational representation of G is a homomorphism

$$\varrho : G \longrightarrow \mathrm{GL}(\mathcal{Z}), \quad \dim_{\mathbb{C}} \mathcal{Z} < \infty,$$

whose components with respect to a basis of the space of all endomorphisms $\mathrm{End}(Z)$ are given by rational functions. Representations are always assumed to be rational. Two representations

$$\varrho : G \longrightarrow \mathrm{GL}(\mathcal{Z})$$
$$\widetilde{\varrho} : G \longrightarrow \mathrm{GL}(\widetilde{\mathcal{Z}})$$

are called isomorphic if there is a vector space isomorphism $\mathcal{Z} \overset{\sim}{\to} \widetilde{\mathcal{Z}}$ which "intertwines" ϱ and $\widetilde{\varrho}$ in the obvious sense. A representation is called irreducible, if $\mathcal{Z}$ is not the zero space and if 0 and $\mathcal{Z}$ are the only invariant subspaces of $\mathcal{Z}$. It is well known that each (rational) representation is isomorphic with a (finite) direct sum of irreducible representations and that the isomorphism classes of these representations are determined uniquely up to their order.

It follows from Schur's lemma that if ϱ is irreducible, there exists an integer k such that

$$\varrho(aE)v = a^k v.$$

We use the notations

$$B(n) = \left\{ \begin{pmatrix} * & \cdots & * \\ \vdots & \ddots & \vdots \\ 0 & \cdots & * \end{pmatrix} \right\}$$

(group of upper triangular matrices),

$$U(n) = \left\{ \begin{pmatrix} 1 & \cdots & * \\ \vdots & \ddots & \vdots \\ 0 & \cdots & 1 \end{pmatrix} \right\}$$

(group of strict upper triangular matrices),

$$B = B(n_1) \times \ldots \times B(n_k),$$
$$U = U(n_1) \times \ldots \times U(n_k).$$

6.5 Definition. *Let*

$$\varrho : G \longrightarrow \mathrm{GL}(\mathcal{Z})$$

be a (rational) representation. A vector of $\mathcal{Z}$ is called a highest weight vector of ϱ if it is invariant under U.

It can be shown that non-zero highest weight vectors always exist if $\mathcal{Z} \neq 0$. More precisely,

6.6 Lemma. *The representation ϱ is irreducible if and only if the space of highest weight vectors is one dimensional.*

Assume that for each $\nu \in \{1, \ldots, k\}$ a rational representation

$$\varrho_\nu : \mathrm{GL}(n_\nu) \longrightarrow \mathrm{GL}(\mathcal{Z}_\nu)$$

is given. Then one constructs in an obvious manner the tensor product $\varrho = \varrho_1 \otimes \ldots \otimes \varrho_k$.

$$\varrho : G \longrightarrow \mathrm{GL}(\mathcal{Z}).$$

The representation space $\mathcal{Z}$ is the tensor product

$$\mathcal{Z} = \mathcal{Z}_1 \otimes \ldots \otimes \mathcal{Z}_k,$$

The group G acts by

$$\varrho(A_1, \ldots, A_k)(a_1 \otimes \ldots \otimes a_k) = \varrho(A_1)a_1 \otimes \ldots \otimes \varrho(A_k)a_k.$$

It is easy to see that ϱ is irreducible if and only if all the components $\varrho_1, \ldots, \varrho_k$ are irreducible. Not quite clear but not difficult is

6.7 Lemma. *Every irreducible rational representation*

$$\varrho : G \longrightarrow \mathrm{GL}(Z)$$

is isomorphic to a tensor product

$$\varrho = \varrho_1 \otimes \ldots \otimes \varrho_k.$$

The components ϱ_ν are irreducible representations, and are uniquely determined up to isomorphism.

Hence it is sufficient to study irreducible representations of a single $G(n)$.

6.8 Proposition. *Let*

$$\varrho : G(n) \longrightarrow \mathrm{GL}(\mathcal{Z})$$

be an irreducible (rational) representation and v a highest weight vector. There exist integers

$$r_1 \geq \ldots \geq r_n$$

such that

$$\varrho \begin{pmatrix} a_1 & \cdots & * \\ \vdots & \ddots & \vdots \\ 0 & \cdots & a_n \end{pmatrix} v = a_1^{r_1} \cdot \ldots \cdot a_n^{r_n} \cdot v.$$

The vector $(r_1, \ldots, r_n)$ is called the **highest weight** of ϱ.

6.9 Theorem. *Two irreducible (rational) representations of $G(n)$ are isomorphic if and only if their highest weights agree. Each vector*

$$(r_1, \ldots, r_n) \in \mathbb{Z}^n; \quad r_1 \geq \ldots \geq r_n,$$

occurs as a highest weight.

Sometimes we denote by

$$[r_1, \ldots, r_n]$$

a fixed representation with highest weight $(r_1, \ldots, r_n)$. In the theory of Siegel modular forms the integer r_n sometimes plays the role of the classical weight. We introduced already in sec.4 the notation

$$k(\varrho) = \text{weight of } \varrho$$

6.10 Remark. *The weight $k(\varrho)$ of ϱ agrees with the last component of the highest weight of ϱ,*

$$r_n = k(\varrho).$$

Examples:

1) The highest weight of the one dimensional representation

$$\varrho(A)v = (\det A)^r v$$

is

$$(r, \ldots, r).$$

2) The tautological representation

$$\varrho : \mathrm{GL}(n, \mathbb{C}) \xrightarrow{\text{id}} \mathrm{GL}(\mathbb{C}^n)$$

is irreducible with highest weight

$$(1, 0, \ldots, 0).$$

An application

Let
$$\varrho : G(n) \longrightarrow \mathrm{GL}(\mathcal{Z})$$
be an irreducible representation with highest weight $(r_1, \ldots r_n)$. For any integer m
$$0 < m < n$$
we consider the subspace
$$\mathcal{Z}_0 \subset \mathcal{Z}$$
consisting of all vectors which are invariant under the abelian group
$$\begin{pmatrix} E^{(m)} & * \\ 0 & E^{(n-m)} \end{pmatrix} .$$

This group is normalized by the subgroup
$$\left\{ \begin{pmatrix} A & 0 \\ 0 & B \end{pmatrix} ; \quad A \in \mathrm{GL}(m); \ B \in \mathrm{GL}(n-m) \right\}.$$

Hence we obtain a representation
$$\widetilde{\varrho} : G(m) \times G(n-m) \longrightarrow \mathrm{GL}(\mathcal{Z}_0),$$
$$\widetilde{\varrho}(A, B)a = \varrho \begin{pmatrix} A & 0 \\ 0 & B \end{pmatrix} a.$$

6.11 Lemma. *The representation*
$$\widetilde{\varrho} : G(m) \times G(n-m) \longrightarrow \mathrm{GL}(\mathcal{Z}_0)$$
is irreducible and isomorphic to
$$[r_1, \ldots, r_m] \otimes [r_{m+1}, \ldots, r_n].$$

Corollary. *Assume that*
$$\varrho : \mathrm{GL}(n, \mathbb{C}) \longrightarrow \mathrm{GL}(\mathcal{Z})$$
is an irreducible representation. Assume furthermore that the subspace
$$\mathcal{Z}_0 \subset \mathcal{Z}$$
of invariants of the group
$$\left\{ \begin{pmatrix} E^{(m)} & * \\ 0 & E^{(n-m)} \end{pmatrix} \right\}$$

contains a non zero vector which is invariant under

$$\varrho \begin{pmatrix} E & 0 \\ 0 & B \end{pmatrix}; \quad B \in \mathrm{SL}(n, \mathbb{Z}).$$

Then the representation

$$\varrho_0 : G(m) \longrightarrow \mathrm{GL}(\mathcal{Z}_0),$$
$$A \longmapsto \varrho \begin{pmatrix} A & 0 \\ 0 & E \end{pmatrix} \Big| \mathcal{Z}_0$$

is irreducible. If $(r_1, \ldots, r_n)$ is the highest weight of ϱ, then $(r_1, \ldots, r_m)$ is the highest weight of ϱ_0. One furthermore has

$$r_{m+1} = \ldots = r_n.$$

II Theta series with polynomial coefficients

1 Coefficient functions and the Gauss transformation

It is well known that theta series with polynomial coefficients as for example

$$\sum_{n_1,\ldots,n_r} P(n_1,\ldots,n_r)\exp\pi i(n_1^2+\ldots+n_r^2)\tau$$

define modular forms if P is a polynomial which satisfies certain conditions of homogenity and harmonicity.

In the theory, which we are going to develop, polynomial coefficients will occur, which are not known to be homogenous or harmonic in advance. Hence it seems to be unavoidable to develop a good deal of the transformation formalism for arbitrary polynomial coefficients. The "inversion formula" will show that in this general context it is necessary to admit also a Z-dependency in the polynomial coefficient. This leads us to the following definition.

1.1 Definition. *A coefficient function of degree n on a vector space $\mathcal{V}$ is a map*

$$P:\mathcal{V}\times\mathbb{H}_n\longrightarrow\mathbb{C},$$

which can be written as

$$P(g,Z)=\sum_{j=1}^{k}P_j(g)A_j(Z),$$

where $P_1,\ldots,P_k$ are polynomials on $\mathcal{V}$ and $A_1,\ldots,A_k$ holomorphic functions on $\mathbb{H}_n$.

We now consider theta series of the type

$$\vartheta_P[\mathbf{m}](Z)=\sum_{g\in\mathbb{Z}^n}P(g+a,Z)\exp\pi i\big\{Z[g+a]+2b'(g+a)\big\},$$

where

$$\mathbf{m}=\begin{bmatrix}a\\b\end{bmatrix};\quad a,b\in\mathbb{C}^n\quad\text{(column vectors)}$$

is the characteristic and

$$P:\mathbb{C}^n\times\mathbb{H}_n\longrightarrow\mathbb{C},$$
$$P(g)=P(g,Z),$$

a coefficient function in the sense of 1.1.

Before we can formulate the inversion formula, we need two further ingredients:

a) The Gauss transformation.

b) A holomorphic matrix root of Z/i.

1.2 Definition. *The Gauss transformed of a polynomial P on $\mathbb{C}^n$ is*

$$P^*(x) = \int_{\mathbb{R}^n} P(x+u)\exp(-\pi u'u)du.$$

Obviously P^* is again a polynomial. If $P(x) = P(x, Z)$ is a coefficient function, then $P^*(x) = P^*(x, Z)$ is a coefficient function too.

1.3 Lemma. *The Gauss transform of a polynomial P is*

$$P^*(x) = e^{\Delta/4\pi}P(x) = \sum \frac{1}{j!}\left(\frac{\Delta}{4\pi}\right)^j P(x).$$

(This sum is finite). Here

$$\Delta = \sum \partial^2/\partial x_\nu^2$$

is the usual Laplacian.

1. Corollary. *The Gauss transformation is invertible,*

$$P = e^{-\Delta/4\pi}P^*.$$

2. Corollary. *If P or P^* is harmonic ($\Delta P = 0$ or $\Delta P^* = 0$), then*

$$P^* = P.$$

3. Corollary. *If P and P^* are both homogeneous polynomials, then they are harmonic.*

Proof. The operator $\exp(\Delta/4\pi)$ is the product of the n operators

$$\exp\left(\frac{\partial^2}{4\pi\partial^2 x_\nu}\right); \quad 1 \le \nu \le n.$$

In the same way the Gauss transform is obtained by applying the Gauss transformation to each variable seperately. Therefore it is sufficient to restrict to the case $n = 1$. In the case $P(x) = x^k$ the formula states

$$\int_{-\infty}^{\infty} (x+u)^k e^{-\pi u^2} du = \sum_{0 \le 2j \le k} \frac{1}{j!}\frac{d^{2j}}{(4\pi)^{2j}d^{2j}x}x^k,$$

which can be proved easily.

Now we introduce a holomorphic matrix root on the Siegel half space.

1.4 Lemma. *There exists a unique holomorphic map*

$$l : \mathbb{H}_n \to \mathcal{Z}_n$$

($\mathcal{Z}_n$ = vector space of all symmetric $n \times n$-matrices) with the properties
a) $e^{l(Z)} = Z/i$,
b) $l(iY)$ is real if $Y > 0$.

Notations:

$$\log(Z/i) = l(Z),$$
$$(Z/i)^{1/2} = \exp\left(\frac{1}{2}\log(Z/i)\right).$$

Proof.
Existence of l: Let $Z \in \mathbb{H}_n$ be a fixed point. All points on the line

$$\alpha(t) := E + t(Z/i - E), \quad 0 \le t \le 1,$$

are invertible matrices because $\mathbb{H}_n$ is convex. We hence may define

$$l(x) = l(x, Z) = \int_0^x \dot{\alpha}(t)/\alpha(t)\,dt, \quad 0 \le x \le 1.$$

The matrices $\alpha(t)$, $\dot{\alpha}(t)$ commute with each other. This implies

$$\dot{l}(t) = \dot{\alpha}(t)/\alpha(t)$$

hence

$$\left[e^{l(t)}/\alpha(t)\right]^{\bullet} = 0.$$

Hence the matrix in the bracket is constant and we obtain

$$e^{l(t)} = \alpha(t)$$

or, in the special case $t = 1$

$$e^{l(1)} = Z/i.$$

Uniqueness of l: It has to be shown that for each $Y > 0$ there exists only one symmetric real matrix A with the property

$$e^A = Y \quad .$$

This well-kown fact can easily be proved by means of an orthogonal transformation of Y into a diagonal matrix. One has to make use of the fact that each matrix which commutes with A commutes with Y too.

2 The general inversion formula

With the notations and definitions introduced in sec.1 we formulate

2.1 Proposition. *We have*

$$\vartheta_P\begin{bmatrix} a \\ b \end{bmatrix}(-Z^{-1}) = \exp(2\pi i a'b)\sqrt{\det(Z/i)}\,\vartheta_Q\begin{bmatrix} -b \\ a \end{bmatrix}(Z),$$

where $Q(g, Z)$ is the Gauss-transform of the polynomial

$$u \longmapsto P((Z/i)^{1/2}u, -Z^{-1})$$

at

$$-i(Z/i)^{1/2}g.$$

Remark. *Even if $P(g) = P(g, Z)$ is a harmonic polynomial independent of Z, the transformed polynomial $Q(g, Z)$ in general depends on Z and is no longer harmonic in g.*

Proof. We proceed as usual and notice that the function

$$a \longmapsto \vartheta_P\begin{bmatrix} a \\ b \end{bmatrix}(Z)$$

is periodic and admits a Fourier expansion

$$\sum_{g \in \mathbb{Z}^n} \alpha(g) \exp 2\pi i g'a.$$

The Fourier coefficients can be computed by means of the Fourier integral. If $Z = iY$ is purely imaginary one obtains

$$\begin{aligned}
\alpha(g) &= \int_0^1 \cdots \int_0^1 \sum_{h \text{ integral}} P(h + a, iY) \\
&\quad \exp\left(-\pi\{Y[h + a] - 2ib'(h + a) + i2g'a\}\right) da \\
&= \int_{-\infty}^\infty \cdots \int_{-\infty}^\infty P(x, iY) \\
&\quad \exp\left(-\pi\{Y[x] - 2i(b - g)'x\}\right) dx \\
&= \exp\left(-\pi Y^{-1}[g - b]\right) \int_{-\infty}^\infty \cdots \int_{-\infty}^\infty P(x, iY) \\
&\quad \exp\left(-\pi\{Y[x + iY^{-1}(g - b)]\}\right) dx \\
&= \exp\left(-\pi Y^{-1}[g - b]\right) \int_{-\infty}^\infty \cdots \int_{-\infty}^\infty P(x - iY^{-1}(g - b), iY) \\
&\quad \exp\left(-\pi Y[x]\right) dx.
\end{aligned}$$

By means of the substitution

$$x = Y^{-1/2}u, \quad dx = \det Y^{-1/2} du$$

we obtain

$$\alpha(g) = \det Y^{-1/2} \exp\left(-\pi Y^{-1}[g-b]\right) \int_{-\infty}^{\infty} \cdots \int_{-\infty}^{\infty}$$
$$P(Y^{-1/2}u - iY^{-1}(g-b), iY) \exp\left(-\pi u'u\right) du.$$

We now have proved 2.1 in the special case $Z = iY^{-1}$. The general case follows by analytic continuation.

We now want to determine the action of arbitrary modular substitutions on the theta series $\vartheta_P[\mathbf{m}]$ and recall the (affine) action of the modular group on characteristics:

$$M\{\mathbf{m}\} = M'^{-1} \cdot \mathbf{m} + \frac{1}{2}\begin{pmatrix}(CD')_0 \\ (AB')_0\end{pmatrix};$$

$$M = \begin{pmatrix} A & B \\ C & D \end{pmatrix} \in \Gamma_n; \quad \mathbf{m} = \begin{pmatrix} a \\ b \end{pmatrix} \in \mathbb{C}^{2n}.$$

Here S_0 denotes the column vector which consists of the diagonal of S. One has

$$MN\{\mathbf{m}\} \equiv M\{N\{\mathbf{m}\}\} \mod 1.$$

2.2 Proposition. *There exists a unique action of the modular group Γ_n on the space of all coefficient functions*

$$(P, M) \longmapsto P_M$$

such that $P_M = P$, if P is constant and such that the transformation formula

$$\vartheta_{P_M}[M\{\mathbf{m}\}](M\langle Z\rangle) = v(M, \mathbf{m}) \det(CZ+D)^{1/2} \vartheta_P[\mathbf{m}](Z)$$

holds for all $M \in \Gamma_n$. Here $v(M, \mathbf{m})$ is a system of complex numbers which is independent of Z and of P.

Proof.
Uniqueness of P_M: The system of numbers $v(M, \mathbf{m})$ is determined by the demand $P_M = P$ for $P \equiv 1$. It is of course the same system considered in I sec.5. The coefficient function of a theta series is uniquely determined by the theta function which is immediately clear if one considers b as variable.

Existence of P_M: If the formula is true for M and N it is also true for $M \cdot N$. This follows from

$$(M \cdot N)\{\mathbf{m}\} \equiv M\{N\{\mathbf{m}\}\} \mod 1$$

and from the fact that a change of the characteristic mod 1 can be absorbed by the coefficient function:

$$\vartheta_{\widetilde{P}}[\widetilde{\mathbf{m}}] = \vartheta_P[\mathbf{m}] \quad \text{if } \widetilde{m} \equiv m \mod 1,$$

where
$$\widetilde{P}(g, Z) = \exp\bigl(2\pi i a'(b - \widetilde{b})\bigr) P(g, Z).$$

It is hence sufficient to prove the transformation formula for generators of the modular group, hence for

a) $Z \longmapsto Z + S;\quad S = S'$ integral,

b) $Z \longmapsto -Z^{-1}$.

The case of a translation is trival. The case of the involution has been treated in 2.1.

In I sec.4 the M-dependency of $v(M, \mathbf{m})$ has been determined. This dependency was very easy to describe if M is contained in the theta group $\Gamma_{n,\vartheta}$, which consists of all $M \in \Gamma_n$ such that
$$(CD')_0 \equiv (AB')_0 \equiv 0 \bmod 2.$$

In this connection it turned out to be useful to use the modified theta series
$$\widetilde{\vartheta}_P[\mathbf{m}](Z) = e^{-\pi i a' b}\vartheta_P[\mathbf{m}](Z).$$

It is also useful to use the linear action
$$(M, \mathbf{m}) \longmapsto M'^{-1}\mathbf{m}$$

instead of the affine action
$$(M, \mathbf{m}) \longmapsto M\{\mathbf{m}\}.$$

For $M \in \Gamma_{n,\vartheta}$ they only differ mod 2. The advantage of the first one is that it defines a real action:
$$(MN)'^{-1}\mathbf{m} = M'^{-1}(N'^{-1}\mathbf{m}).$$

Combining the results of I sec.5 with proposition 4.1 we obtain

2.4 Proposition. *There exists a unique action of the theta group $\Gamma_{n,\vartheta}$ on the space of all coefficient functions*
$$(P, M) \longmapsto P^M,$$

such that $P^M = P$ if P is constant and such that the formula
$$\widetilde{\vartheta}_{P^M}[M'^{-1}\mathbf{m}](M\langle Z\rangle) = v_\vartheta(M) \det(CZ + D)^{1/2}\widetilde{\vartheta}_P[\mathbf{m}](Z)$$

holds for all $M \in \Gamma_{n,\vartheta}$. Here v_ϑ is the theta multiplier system considered in I sec.5.. We have
$$P^{MN} = \bigl(P^N\bigr)^M.$$

In the next section we will determine an explicit formula for the action $(P, M) \mapsto P^M$.

3   The action of the theta group on coefficient functions

We want to derive an explicit formula for the action (s. sec.2)

$$(P, M) \longmapsto P^M \quad (M \in \Gamma_{n,\vartheta}).$$

We restrict to the case of an invertible D. This is not necessary but simplifies the calculations. It is in some sense enough because of the following

3.1 Remark. *Each congruence subgroup $\Gamma \subset \mathrm{Sp}(n, \mathbb{R})$ is generated by the subset of all M with invertible D.*

Proof. If we multiply a matrix

$$M = \begin{pmatrix} A & B \\ C & D \end{pmatrix} \in \Gamma$$

from the right by a translation matrix

$$\begin{pmatrix} E & S \\ 0 & E \end{pmatrix},$$

the matrix D changes into

$$D + CS.$$

It is therefore sufficient to find a S such that

$$\det(D + CS) \neq 0.$$

Because S may run through a complete lattice, we may consider it as a variable, i.e. we can admit arbitrary real S. The construction is not difficult. We only give the hint that one has to make use of the fact that CD' is symmetric (because one is restricted to symmetric S, actually $S = tE$ will be sufficient).

The determination of the explicit formula can be considered as a generalization of the calculation, which expressed the theta multiplier system $v_\vartheta(M)$ as a Gauss sum (I sec.5). In fact the following calculation will also give a new proof of that formula. Again we start with the formula

$$M\langle Z \rangle = W + R,$$
$$W = D'^{-1} Z (CZ + D)^{-1}; \quad R = BD^{-1}.$$

We use the notation

$$M'^{-1} \mathbf{m} = \mathbf{n} = \begin{pmatrix} \alpha \\ \beta \end{pmatrix}.$$

One has

$$\vartheta_{PM}[M'^{-1}\mathbf{m}](M\langle Z \rangle)$$
$$= \sum P^M(g + \alpha, M\langle Z \rangle) \exp \pi i \big(W[g + \alpha] + R[g + \alpha] + 2\beta'(g + \alpha) \big)$$
$$= \exp \pi i \big(R[\alpha] + 2\beta'\alpha \big) \cdot$$
$$\qquad \sum P^M(g + \alpha, M\langle Z \rangle) \exp \pi i \big(W[g + \alpha] + R[g] + 2(R\alpha + \beta)'g \big).$$

We want to simplify this formula and assume

$$R\alpha = -\beta,$$

which will be sufficient for our purpose. Then one obtains

$$\vartheta_{PM}[M'^{-1}\mathbf{m}](M\langle Z\rangle)\exp\pi iR[\alpha]$$
$$= \sum P^M(g+\alpha, M\langle Z\rangle)\exp\pi i\big(W[g+\alpha] + R[g]\big).$$

We notice that

$$\exp\pi iR[g]$$

remains unchanged if one replaces

$$g \longmapsto g + Dh.$$

(One has to make use of the fact that

$$BD^{-1}[h] \equiv 0 \bmod 2,$$

which is true because M is in the theta group.)

If g_0 runs through a system of representatives of $\mathbb{Z}^n/D\mathbb{Z}^n$, one obtains for the above sum

$$\vartheta_{PM}[M'^{-1}\mathbf{m}](M\langle Z\rangle)\exp\pi iR[\alpha] =$$
$$\sum_{g_0 \bmod D} \exp\pi iR[g_0] \sum_{g \text{ integer}} P^M(g_0 + Dg + \alpha, M\langle Z\rangle)\cdot\exp\pi iW[g_0 + Dg + \alpha].$$

If we put

$$\widetilde{\alpha} = D^{-1}(\alpha + g_0), \quad \widetilde{\beta} = 0,$$
$$\widetilde{\mathbf{n}} = \begin{pmatrix}\widetilde{\alpha}\\\widetilde{\beta}\end{pmatrix},$$
$$\widetilde{P}(g, Z) = P^M(Dg, Z[D^{-1}] + R),$$

we obtain

$$\vartheta_{\widetilde{P}}[\widetilde{\mathbf{n}}](W[D])$$
$$= \sum \widetilde{P}(g+\widetilde{\alpha}, W[D])\exp\pi iW[D][g+\widetilde{\alpha}]$$
$$= P^M(Dg+\alpha+g_0, W+R)\exp\pi iW[Dg+\alpha+g_0]$$

and therefore

$$\vartheta_{PM}[M'^{-1}\mathbf{m}](M\langle Z\rangle)\exp\pi iR[\alpha]$$
$$= \sum_{g_0 \bmod D} \exp\big(\pi iR[g_0]\big)\vartheta_{\widetilde{P}}[\widetilde{\mathbf{n}}](W[D]).$$

Now we apply the inversion formula:

$$\vartheta_{\widetilde{P}}[\widetilde{\mathbf{n}}](W[D]) = \sqrt{\det\big(-W[D]^{-1}/i\big)}\,\vartheta_{\widetilde{Q}}\begin{bmatrix}-\widetilde{\beta}\\\widetilde{\alpha}\end{bmatrix}(-W[D]^{-1}),$$

where $\widetilde{Q}(g, Z)$ is the Gauss transform of the polynomial

$$\widetilde{P}\big((Z/i)^{1/2}u, -Z^{-1}\big) =$$
$$P^M\big(D(Z/i)^{1/2}u, -Z^{-1}[D^{-1}] + R\big)$$

evaluated at

$$-i(Z/i)^{1/2}g.$$

What we have proved is

$$\vartheta_{P^M}[M'^{-1}\mathbf{m}](M\langle Z\rangle)\exp \pi iR[\alpha] =$$
$$\sqrt{\det\big(-W[D]^{-1}/i\big)} \sum_{g_0 \bmod D} \exp\big(\pi iR[g_0]\big) \cdot \vartheta_{\widetilde{Q}}\begin{bmatrix} 0 \\ \widetilde{\alpha} \end{bmatrix}\big(-W[D]^{-1}\big).$$

We now have replaced in the transformation formula in 2.4 $M\langle Z\rangle$ on the left hand side by

$$-W[D]^{-1} = -Z^{-1} - D^{-1}C.$$

The same will be done on the right hand side. Because of

$$\mathbf{m} = \begin{pmatrix} a \\ b \end{pmatrix} = M'\mathbf{n} = \begin{pmatrix} A'\alpha + C'\beta \\ B'\alpha + D'\beta \end{pmatrix}$$

and because of

$$\beta = -R\alpha$$

we have

$$a = D^{-1}\alpha, \quad b = 0.$$

This implies

$$\widetilde{\vartheta}_P[\mathbf{m}](Z) = \vartheta_P[\mathbf{m}](Z) = \sqrt{\det(Z/i)^{-1}}\vartheta_Q\begin{bmatrix} -b \\ a \end{bmatrix}(-Z)^{-1}.$$

If we make use of the formula

$$\widetilde{\vartheta}_{P^M}[M'^{-1}\mathbf{m}] = \exp\big(\pi iR[\widetilde{\alpha}]\big)\vartheta_{P^M}[M'^{-1}\mathbf{m}],$$

the transformation formula in 2.4 may be rewritten as

$$\sqrt{\det\big((-Z^{-1} - D^{-1}C)/i\big)} \sum_{g_0 \bmod D} \exp\big(\pi iR[g_0]\big)\widetilde{\vartheta}_{\widetilde{Q}}\begin{bmatrix} 0 \\ \widetilde{\alpha} \end{bmatrix}(-Z^{-1} - D^{-1}C)$$
$$= v_\vartheta(M)\sqrt{\det(CZ + D)}\sqrt{\det(Z/i)^{-1}}\widetilde{\vartheta}_Q\begin{bmatrix} b \\ -a \end{bmatrix}(-Z^{-1}),$$

or explicitly

$$\sqrt{\det\bigl(-(Z^{-1}+D^{-1}C)/i\bigr)} \sum_{g_0 \bmod D} \exp \pi i R[g_0] \cdot$$

$$\sum_{g \text{ integral}} \tilde{Q}(g, -Z^{-1} - D^{-1}C) \cdot$$

$$\exp \pi i \bigl(-Z^{-1}[g] - D^{-1}C[g] + 2g'D^{-1}(\alpha + g_0)\bigr)$$

$$= v_\vartheta(M)\sqrt{\det(CZ+D)}\sqrt{\det(Z/i)^{-1}} \cdot$$

$$\sum_{g \text{ integral}} Q(g, -Z^{-1}) \exp \pi i \bigl(-Z^{-1}[g] + 2g'D^{-1}\alpha\bigr).$$

This is an identity between Fourier series with respect to the variable α. Comparison of the Fourier coefficients gives for each g:

$$\sqrt{\det\bigl(-(Z^{-1}-D^{-1}C)/i\bigr)} \sum_{g_0 \bmod D} \exp \pi i R[g_0]$$

$$\tilde{Q}(g, -Z^{-1} - D^{-1}C) \exp \pi i \bigl(-D^{-1}C[g] + 2g'D^{-1}g_0\bigr)$$

$$= v_\vartheta(M)\sqrt{\det(CZ+D)}\sqrt{\det(Z/i)^{-1}} \cdot Q(g, -Z^{-1}).$$

Using the conventions about the choice of the square root made in I sec.5 we obtain

$$\tilde{Q}(g, -Z^{-1} - D^{-1}C) \cdot$$

$$(\det D)^{-1/2} \sum_{g_0 \bmod D} \exp \pi i \bigl(R[g_0] - D^{-1}C[g] - 2g'D^{-1}g_0\bigr)$$

$$= v_\vartheta(M) \cdot Q(g, -Z^{-1}).$$

In the special case $P(g, Z) = 1$ one has

$$\tilde{P}(g, Z) = \tilde{Q}(g, Z) = 1$$

and

$$P^M(g, Z) = Q^M(g, Z) = 1.$$

We hence obtain a formula for the multiplier system

$$v_\vartheta(M) = (\det D)^{-1/2} \cdot \sum_{g_0 \bmod D} \exp \pi i \bigl(R[g_0] - D^{-1}C[g] - 2g'D^{-1}g_0\bigr).$$

This sum is especially independent of g and hence has to be computed only for special g, for example $g = 0$. We recover the formula for the theta multiplier system, which we derived in I sec.5 (5.8). A second application now gives us the desired formula for the action of the theta group on coefficient functions.

$$\tilde{Q}(g, -Z^{-1} - D^{-1}C) = Q(g, -Z^{-1}).$$

3.2 Proposition. *The action of the theta group on the coefficient functions can be described as follows.*

Let $\widetilde{Q}(g, Z)$ *be the Gauss transform of*

$$u \longmapsto P^M\big(D \cdot (Z/i)^{1/2}u, -Z^{-1}[D^{-1}] + BD^{-1}\big)$$

at

$$-i(Z/i)^{1/2}g$$

and let $Q(g, Z)$ *be the Gauss transform of*

$$u \longmapsto P\big((Z/i)^{1/2}u, -Z^{-1}\big)$$

at

$$-i(Z/i)^{1/2}g.$$

Then

$$Q(g, Z) = \widetilde{Q}(g, Z - D^{-1}C).$$

This proposition gives in fact an explicit formula (which we do not write down), because the Gauss transformation is invertible.

4   The Eichler imbedding

In the next section we will introduce theta series attached to positive definite quadratic forms. They can be considered as a generalization of the series introduced in sec.1. But they can also be obtained by specialization from those special ones. This was discovered by M. EICHLER and is known as the **Eichler imbedding trick**. In this section we define the Eichler imbedding.

The Kronecker product of two matrices

$$A = A^{(m,n)}, \quad B = B^{(r,s)}$$

is the (mr, ns)-matrix defined by

$$A \otimes B = \begin{pmatrix} Ab_{11} & \dots & Ab_{1s} \\ \vdots & & \vdots \\ Ab_{r1} & \dots & Ab_{rs} \end{pmatrix}.$$

4.1 Remark. *The Kronecker product is associative and bilinear. Furthermore the following formulae are easily verified (under obvious assumptions about the size of the matrices)*

$$(A_1 \otimes B_1)(A_2 \otimes B_2) = (A_1 A_2) \otimes (B_1 B_2),$$
$$(A \otimes B)^{-1} = A^{-1} \otimes B^{-1},$$
$$\det(A \otimes B) = (\det A)^n \cdot (\det B)^m \quad (A = A^{(m)}, \ B = B^{(n)}),$$
$$\sigma(A \otimes B) = \sigma(A)\sigma(B) \quad (\sigma \text{ denotes the trace}),$$
$$(A \otimes B)' = A' \otimes B'.$$

The proof of this and the following remark is very easy and can be left to the reader.

4.2 Remark. *If $A = A^{(m)}$ and $B = B^{(n)}$ are symmetric matrices, one has*

$$(A \otimes B)[g] = \sigma(A[G]B),$$

where

$$G = G^{(m,n)} = (g_1, \ldots, g_n)$$

denotes a $m \times n$-matrix and g the column vector

$$g = \begin{pmatrix} g_1 \\ \vdots \\ g_n \end{pmatrix} .$$

Corollary. *If $S = S^{(r)}$, $Y = Y^{(n)}$ are real symmetric positive definite matrices, then $S \otimes Y$ is also symmetric and positive definite.*

We especially obtain an imbedding of Siegel half spaces

$$\mathbb{H}_n \longrightarrow \mathbb{H}_{nr},$$
$$Z \longmapsto S \otimes Z.$$

4.3 Remark. *The imbedding $Z \longmapsto S \otimes Z$ is compatible with the action of the symplectic groups in the following sense. The map*

$$\mathrm{Sp}(n, \mathbb{R}) \longrightarrow \mathrm{Sp}(nr, \mathbb{R})$$
$$M \longmapsto M^S$$
$$\| \qquad\qquad \|$$
$$\begin{pmatrix} A & B \\ C & D \end{pmatrix} \quad \begin{pmatrix} E \otimes A & S \otimes B \\ S^{-1} \otimes C & E \otimes D \end{pmatrix}$$

is an injective homomorphism with the property

$$M^S \langle S \otimes Z \rangle = S \otimes (M \langle Z \rangle).$$

The proof is an immediate consequence of the rules 4.1. We shall now define a certain important subgroup of the symplectic group.

4.4 Definition. *For a symmetric positive definite rational matrix $S = S^{(r)}$ we define*

$$\Gamma_n(S) = \{ M \in \mathrm{Sp}(n, \mathbb{R}), \ M^S \in \Gamma_{n,\vartheta} \}.$$

Because S is rational, this group is obviously a congruence group.

We recall that S is called **even**, if S is integral and if the diagonal elements of S are even, equivalently

$$S[g] \equiv 0 \bmod 2 \quad \text{for } g \in \mathbb{Z}^r.$$

4.5 Remark. *Assume that S is even and that q is a natural number such that qS^{-1} is even too. Then $\Gamma_n(S)$ contains the group $\Gamma_{n,0}[q]$. (s. I sec.1)*

4.6 Remark. *Let q be a natural number such that*

$$qS \text{ and } qS^{-1}$$

both are integral. Then $\Gamma_n(S)$ contains Igusa's group $\Gamma_n[q, 2q]$. (s. I sec.1)

The proofs are trivial.

In the next section we need the isomorphism

$$\mathbb{C}^{(r,n)} \longrightarrow \mathbb{C}^{rn}$$

$$G = (g_1, \ldots, g_n) \longmapsto g = \begin{pmatrix} g_1 \\ \vdots \\ g_n \end{pmatrix},$$

which attaches to the matrix G, the big column g built from the columns of G (compare 4.2). We say that the matrix G and the big vector g correspond to each other and we denote this correspondence symbolically by

$$G \longleftrightarrow g.$$

4.7 Remark. *Let*

$$A \in \mathbb{C}^{(r,r)} \quad B \in \mathbb{C}^{(n,n)}$$

be two square matrices and

$$G \in \mathbb{C}^{(r,n)}$$

an $r \times n$-matrix. Then the corresponding big vector to the matrix AGB' is $(A \otimes B)g$, where g denotes the corresponding vector of G:

$$AGB' \longleftrightarrow (A \otimes B)g.$$

5   Theta series with respect to positive definite quadratic forms

We consider (generalized) theta series of the type

$$\vartheta_P[\mathbf{m}](S; Z) =$$
$$\sum_{G \in \mathbb{Z}^{(r,n)}} P\big(S^{1/2}(G+U), Z\big) \exp \pi i \sigma\big(S[G+U]Z + 2V'(G+U)\big)$$

and also the modified series

$$\widetilde{\vartheta}_P[\mathbf{m}](S; Z) = \exp\big(-\pi i \sigma(U'V)\big)\vartheta_P[\mathbf{m}](S; Z).$$

Here we have used the following notations:

1) $S \in \mathbb{R}^{(r,r)}$ is a real symmetric positive definite matrix.

2) $P : \mathbb{C}^{(r,n)} \times \mathbb{H}_n \longrightarrow \mathbb{C}$ is a coefficient function in the sense of 1.1.

3)   The "characteristic"

$$\mathbf{m} = \begin{pmatrix} U \\ V \end{pmatrix}; \quad U, V \in \mathbb{C}^{(r,n)},$$

is a pair of complex $r \times n$-matrices and Z varies in the Siegel upper half space of degree n.

In the special case of the 1×1-matrix

$$S = (1)$$

we recover the theta series considered in sec.1–3, more precisely

$$\vartheta_P\begin{bmatrix} a' \\ b' \end{bmatrix}((1); Z) = \vartheta_P\begin{bmatrix} a \\ b \end{bmatrix}(Z);$$
$$\widetilde{\vartheta}_P\begin{bmatrix} a' \\ b' \end{bmatrix}((1); Z) = \widetilde{\vartheta}_P\begin{bmatrix} a \\ b \end{bmatrix}(Z).$$

It is a very remarkable fact that conversely the generalized theta series can be obtained through specialization from the old ones. This is the so-called Eichler imbedding trick. We use the notations of sec.4.

5.2 Remark. *For each holomorphic function*

$$A : \mathbb{H}_n \longrightarrow \mathbb{C}$$

there exists a holomorphic function

$$B : \mathbb{H}_{rn} \longrightarrow \mathbb{C}$$

such that

$$B(S \otimes Z) = A(Z).$$

Proof. Let

$$
\widetilde{Z} = \begin{pmatrix} S^{1/2} Z_{11} S^{1/2} & \cdots & S^{1/2} Z_{1n} S^{1/2} \\ \vdots & \ddots & \vdots \\ S^{1/2} Z_{n1} S^{1/2} & \cdots & S^{1/2} Z_{nn} S^{1/2} \end{pmatrix}
$$

$$
= \begin{pmatrix} S^{1/2} & \cdots & 0 \\ \vdots & \ddots & \vdots \\ 0 & \cdots & S^{1/2} \end{pmatrix} \begin{pmatrix} Z_{11} & \cdots & Z_{1n} \\ \vdots & \ddots & \vdots \\ Z_{n1} & \cdots & Z_{nn} \end{pmatrix} \begin{pmatrix} S^{1/2} & \cdots & 0 \\ \vdots & \ddots & \vdots \\ 0 & \cdots & S^{1/2} \end{pmatrix}
$$

where $Z_{ij} \in \mathbb{C}^{(r,r)}$. We denote by

$$
z_{ij} := \left(Z_{ij} \right)_{11} \qquad (1 \le i, j \le r)
$$

the 1×1-component of the matrix Z_{ij}. The matrix

$$
Z = (z_{ij})
$$

is contained in $\mathbb{H}_n$. If we define

$$
B(\widetilde{Z}) = A(Z),
$$

we obtain

$$
B(S \otimes Z) = B \begin{pmatrix} S z_{11} E & \cdots & S z_{1n} E \\ \vdots & \ddots & \vdots \\ S z_{n1} E & \cdots & S z_{nn} E \end{pmatrix} = A(Z).
$$

From 5.2 follows

5.3 Remark. *For each coefficient function*

$$
P : \mathbb{C}^{(r,n)} \times \mathbb{H}_n \longrightarrow \mathbb{C}
$$

there exists a coefficient function

$$
P_0 : \mathbb{C}^{rn} \times \mathbb{H}_{rn} \longrightarrow \mathbb{C},
$$

such that

$$
P_0(g, S \otimes Z) = P(S^{1/2} G, Z),
$$

where G denotes the $r \times n$-matrix associated with the big column g (s. sec.4).

5.4 Remark. *Let a, b be the big columns which correspond to U, V in the sense of sec.4,*

$$
a \longleftrightarrow U; \quad b \longleftrightarrow V.
$$

Furthermore let P_0 be a coefficient function on $\mathbb{C}^{rn} \times \mathbb{H}_{rn}$, which is connected with P as in 5.3:

$$
P_0(g, S \otimes Z) = P(S^{1/2} G, Z),
$$
$$
(g \longleftrightarrow G).
$$

Then we have

$$
\vartheta_P \begin{bmatrix} U \\ V \end{bmatrix} (S; Z) = \vartheta_{P_0} \begin{bmatrix} a \\ b \end{bmatrix} (S \otimes Z);
$$

$$
\widetilde{\vartheta}_P \begin{bmatrix} U \\ V \end{bmatrix} (S; Z) = \widetilde{\vartheta}_{P_0} \begin{bmatrix} a \\ b \end{bmatrix} (S \otimes Z).
$$

The proof is trivial.

It remains to reformulate the theta transformation formula. We start with the inversion formula. From the inversion formula 2.1 we obtain

$$\vartheta_P \begin{bmatrix} U \\ V \end{bmatrix} (S^{-1}, -Z^{-1}) = \vartheta_{P_0} \begin{bmatrix} a \\ b \end{bmatrix} \left(-(S \otimes Z)^{-1} \right)$$

$$= \exp(2\pi i a' b) \sqrt{\det(S \otimes Z/i)} \, \vartheta_{Q_0} \begin{bmatrix} -b \\ a \end{bmatrix} (S \otimes Z),$$

where $Q_0(g, \widetilde{Z})$ denotes the Gauss transform of the polynomial

$$u \longmapsto P_0 \left((\widetilde{Z}/i)^{1/2} u, -\widetilde{Z}^{-1} \right)$$

at

$$-i(\widetilde{Z}/i)^{1/2} g.$$

If we specialize $\widetilde{Z} = S^{-1} \otimes Z$, we obtain that $Q_0(g, S \otimes Z)$ is the Gauss transform of

$$U \longrightarrow P\left(U(Z/i)^{1/2}, -Z^{-1} \right)$$

at

$$-iS^{-1/2} G(Z/i)^{1/2}.$$

$\big($In this connection U denotes a variable $r \times n$-matrix and not a characteristic. The Gauss transform is of course defined by

$$P^*(X) = \int_{\mathbb{R}^{(r,n)}} P(X + U) \exp\left(-\pi \sigma(U'U) \right) dU.\big)$$

We define

$$Q(S^{1/2} G, Z) = Q_0(g, S \otimes Z)$$

and obtain

5.5 Proposition. *We have*

$$\vartheta_P \begin{bmatrix} U \\ V \end{bmatrix} (S^{-1}; -Z^{-1}) = \exp\left(2\pi i \sigma(U'V) \right) (\det S)^{n/2}$$

$$\det(Z/i)^{r/2} \vartheta_Q \begin{bmatrix} -V \\ U \end{bmatrix} (S; Z),$$

where $Q(G, Z)$ is the Gauss transform of the polynomial

$$U \longmapsto P(U(Z/i)^{1/2}, -Z^{-1})$$

at

$$-iG(Z/i)^{1/2}.$$

In a similar way we generalize the transformation formulae 2.4 and 3.2.

5.6 Theorem. *Let $S \in \mathbb{Q}^{(r,r)}$ be a rational symmetric positive matrix. There exists an action of the group $\Gamma_n(S)$ (4.4)*

$$(P, M) \longmapsto P^M$$

on the space of coefficient functions

$$P : \mathbb{C}^{(r,n)} \times \mathbb{H}_n \longrightarrow \mathbb{C},$$

such that $P^M = P$ if P is constant and such that the formula

$$\tilde{\vartheta}_{PM} \begin{bmatrix} \tilde{U} \\ \tilde{V} \end{bmatrix} (S; M\langle Z\rangle) = v_S(M)\det(CZ + D)^{r/2}\tilde{\vartheta}_P \begin{bmatrix} U \\ V \end{bmatrix}(S; Z)$$

is valid for all $M \in \Gamma_n(S)$.

Here we use the notations

$$\tilde{U} = UD' - S^{-1}VC', \quad \tilde{V} = -SUB' + VA'; \qquad M = \begin{pmatrix} A & B \\ C & D \end{pmatrix},$$

and

$$v_S(M) := v_\vartheta(M^S)$$
$$\left(= \sqrt{\det D}^r \sum_{G \in \mathbb{Z}^{(r,n)}} \exp \pi i\sigma(BD^{-1}S[G]) \quad \text{if } \det D \neq 0 \right).$$

The action on the coefficient functions can be described explicitely as follows ($\det D \neq 0$):

Let $\tilde{Q}(G, Z)$ be the Gauss transform of

$$U \longmapsto P^M\left(U(Z/i)^{1/2}D', -Z^{-1}[D^{-1}] + BD^{-1}\right)$$

at

$$-iG(Z/i)^{1/2}$$

and let $Q(G, Z)$ be the Gauss transform of

$$U \longmapsto P\left(U(Z/i)^{1/2}, -Z^{-1}\right)$$

at

$$-iG(Z/i)^{1/2}.$$

Then

$$Q(G, Z) = \tilde{Q}(G, Z - D^{-1}C).$$

Proof. We choose a coefficient function

$$P_0 : \mathbb{C}^{rn} \times \mathbb{H}_{rn} \longrightarrow \mathbb{C},$$

such that (5.3)
$$P_0(g, S \otimes Z) = P(S^{1/2}G, Z).$$

After that we define

$$Q_0(g, \widetilde{Z}) \quad (g \in \mathbb{C}^{rn}, \ \widetilde{Z} \in \mathbb{H}_{rn})$$

to be the Gauss transform of

$$u \longmapsto P_0\big((\widetilde{Z}/i)^{1/2}u, -\widetilde{Z}^{-1}\big)$$

at

$$-i(\widetilde{Z}/i)^{1/2}g.$$

As in 3.2 we define for each

$$\widetilde{M} = \begin{pmatrix} \widetilde{A} & \widetilde{B} \\ \widetilde{C} & \widetilde{D} \end{pmatrix} \in \Gamma_{rn,\vartheta}$$

the coefficient function $\widetilde{Q}_0$ by

$$\widetilde{Q}_0(g, \widetilde{Z}) = Q_0(g, \widetilde{Z} + \widetilde{D}^{-1}\widetilde{C}).$$

Finally $P_0^{\widetilde{M}}$ is defined such that $\widetilde{Q}_0(g, \widetilde{Z})$ is the Gauss transform of

$$u \longmapsto P_0^{\widetilde{M}}\big(\widetilde{D} \cdot (\widetilde{Z}/i)^{1/2}u, -\widetilde{Z}^{-1}[\widetilde{D}^{-1}] + \widetilde{B}\widetilde{D}^{-1}\big)$$

at

$$-i(\widetilde{Z}/i)^{1/2}g.$$

We specialize these formulae to the case

$$\widetilde{M} = M^S = \begin{pmatrix} E \otimes A & S \otimes B \\ S^{-1} \otimes C & E \otimes D \end{pmatrix}; \quad M \in \Gamma_n(S)$$

and

$$\widetilde{Z} = S^{-1} \otimes Z.$$

With the notations

$$Q(S^{1/2}G, Z) = Q_0(g, S \otimes Z),$$
$$\widetilde{Q}(S^{1/2}G, Z) = \widetilde{Q}_0(g, S \otimes Z),$$
$$P^M(S^{1/2}G, Z) = P_0^{M^S}(g, S \otimes Z),$$

we obtain:

1) $Q(G, Z)$ is the Gauss transform of

$$U \longmapsto P\big(U(Z/i)^{1/2}, -Z^{-1}\big)$$

at

$$-iG(Z/i)^{1/2}.$$

2) $\widetilde{Q}(G, Z)$ is the Gauss transform of

$$U \longmapsto P^M\big((U(Z/i)^{1/2}D', -Z^{-1}[D^{-1}] + BD^{-1}\big)$$

at

$$-iG(Z/i)^{1/2}.$$

The defining equality of P^M reads

$$Q(G, Z) = \widetilde{Q}(Z - D^{-1}C).$$

Now we are able to specialize formula 3.2

$$\widetilde{\vartheta}_{P_0^{\widetilde{M}}}\left[\widetilde{M}'^{-1}\begin{pmatrix} a \\ b \end{pmatrix}\right](\widetilde{M}\langle\widetilde{Z}\rangle) = v_\vartheta(\widetilde{M})\det(\widetilde{C}\widetilde{Z} + \widetilde{D})^{1/2}\widetilde{\vartheta}_{P_0}\begin{bmatrix} b \\ -a \end{bmatrix}(\widetilde{Z})$$

to the case

$$\widetilde{M} = M^S = \begin{pmatrix} E \otimes A & S \otimes B \\ S^{-1} \otimes C & E \otimes D \end{pmatrix}; \quad M \in \Gamma_n(S)$$

and

$$\widetilde{Z} = S \otimes Z.$$

The characteristic $(M^S)'^{-1}\begin{pmatrix} a \\ b \end{pmatrix}$ equals

$$\begin{pmatrix} (E \otimes B)a - (S^{-1} \otimes C)b \\ -(S \otimes B)a + (E \otimes A)b \end{pmatrix} \longleftrightarrow \begin{pmatrix} UB' - S^{-1}VC' \\ -SUB' + VA' \end{pmatrix}.$$

The simple observation

$$\det\big((S^{-1} \otimes C)(S \otimes Z) + (E \otimes D)^{1/2}\big) = \det(CZ + D)^{r/2}$$

completes the proof of theorem 5.6.

6   Theta series with harmonic forms as coefficients

The theta transformation formalism simplifies considerably, if the polynomial coefficients are harmonic forms.

A function

$$P : \mathbb{C}^{(r,n)} \longrightarrow \mathbb{C}$$

is called **harmonic** if it is harmonic in the rn variables in the usual sense, i.e.

$$\sum_{ik} \frac{\partial^2 P(X)}{(\partial x_{ik})^2} = 0.$$

6.1 Definition. *A polynomial*

$$P : \mathbb{C}^{(r,n)} \longrightarrow \mathbb{C}$$

is called pluriharmonic *if the polynomials*

$$X \longmapsto P(XA)$$

are harmonic for all $A \in \mathbb{C}^{(n,n)}$.

The following two remarks are very easy to prove.

6.2 Remark. *A polynomial P is pluriharmonic if and only if it satisfies the system of differential equations*

$$\sum_{\nu=1}^{r} \frac{\partial^2 P(X)}{\partial x_{\nu i} \partial x_{\nu k}} = 0 \quad for \quad 1 \leq i, k \leq n.$$

We denote by

$$O(r, \mathbb{C}) = \{ A \in \mathrm{GL}(r, \mathbb{C}); \quad A'A = E \}$$

the orthogonal group.

6.3 Remark. *The group*

$$O(r, \mathbb{C}) \times \mathrm{GL}(n, \mathbb{C})$$

acts on the space of pluriharmonic forms on $\mathbb{C}^{(r,n)}$ by means of

$$P(X) \longmapsto P(B^{-1} X A).$$

Let S be any rational symmetric positive $r \times r$-matrix. The formula for the action of the group $\Gamma_n(S)$ on coefficient functions P simplifies considerably if $G \mapsto P(G, Z)$ is a pluriharmonic polynomial in G. In this case

$$U \longmapsto P\big(U(Z/i)^{1/2}, *\big)$$

is a harmonic polynomial and therefore agrees with its Gauss transform. This gives us (with the notations of 5.6)

$$Q(G, Z) = P(-GZ, -Z^{-1})$$

and

$$\widetilde{Q}(G, Z) = P\big(-G(Z + D^{-1}C), -(Z + D^{-1}Z)^{-1}\big).$$

The inversion formula for the Gauss transformation (1.3, 1.Corollary) shows that $G \mapsto P^M(G, Z)$ is harmonic too. This implies

$$\widetilde{Q}(G, Z) = P^M\big(-GZD', -Z^{-1}[D^{-1}] + BD^{-1}\big)$$

In the expression of P^M as function of P, the Gauss transformation disappeared! What we obtained is

$$P^M\big(-GZD', -Z^{-1}[D^{-1}] + BD^{-1}\big) = P\big(-G(Z + D^{-1}C), -(Z + D^{-1}C)^{-1}\big),$$

equivalently

6.4 Proposition. *The action of the group $\Gamma_n(S)$ (s. 5.6) on a pluriharmonic coefficient function $P(G^{(r,n)}, Z)$ is given by the formula*

$$P^M(G, M\langle Z\rangle) = P\big(G(CZ + D)'^{-1}, Z\big).$$

Proof. The statement is equivalent to

$$P^M(G, Z) = P\big(G(CM^{-1}\langle Z\rangle + D)'^{-1}, M^{-1}\langle Z\rangle\big)$$
$$= P\big(G(A - ZC)'M^{-1}\langle Z\rangle\big)$$

or

$$P^M\big(-GZD', -Z^{-1}[D^{-1}] + BD^{-1}\big)$$
$$= P\big(-GZD'(A + Z^{-1}[D^{-1}]C - BD^{-1}C), M^{-1}\langle -Z^{-1}[D^{-1}] + BD^{-1}\rangle\big).$$

Proposition 6.4 is therefore equivalent to

a) $$ZD'(A + Z[D^{-1}]C - BD^{-1}C) = Z + D^{-1}C,$$
b) $$M^{-1}\langle -Z^{-1}[D^{-1}] + BD^{-1}\rangle = -(Z + D^{-1}C)^{-1}.$$

Both equalities are easy consequences of the symplectic relations 1.1.

We observe that the transformed polynomial P^M may depend on Z even if P does not. The following consideration will remedy this situation.

If P is a fixed polynomial on $\mathbb{C}^{(r,n)}$, then all the polynomials

$$G \longmapsto P(GA'), \quad A \in \mathrm{GL}(n,\mathbb{C}),$$

generate a finite dimensional vector space of polynomials. If $P_1, \ldots, P_m$ denotes a basis of this vector space, we have a transformation formula

$$\begin{pmatrix} P_1(GA') \\ \vdots \\ P_m(GA') \end{pmatrix} = \varrho_0(A) \cdot \begin{pmatrix} P_1(G) \\ \vdots \\ P_m(G) \end{pmatrix},$$

where

$$\varrho_0 : \mathrm{GL}(n,\mathbb{C}) \longrightarrow \mathrm{GL}(m,\mathbb{C})$$

is obviously a polynomial representation. This leads us to consider vector valued polynomials.

6.5 Definition. *Let $\mathcal{Z}$ be a finite dimensional complex vector space and*

$$\varrho : \mathrm{GL}(n,\mathbb{C}) \longrightarrow \mathrm{GL}(\mathcal{Z})$$

be a polynomial representation. A harmonic form with respect to ϱ is a polynomial function

$$P : \mathbb{C}^{(r,n)} \longrightarrow \mathcal{Z}$$

with the following two properties.

1) $P(GA') = \varrho_0(A)P(G)$ for all $A \in \mathrm{GL}(n,\mathbb{C})$.

2) P is harmonic ($\Delta P = 0$).

(Operators like the Laplacian or the Gauss transformation apply to vector valued polynomials componentwise with respect to a basis of $\mathcal{Z}$.) Each harmonic form is of course pluriharmonic. The above consideration shows conversely:

6.6 Remark. *Each pluriharmonic polynomial is the component of a harmonic form*

We can say even a little more:

The determination of all harmonic forms is equivalent to the description of the representation

$$P(G) \longmapsto P(GA); \quad A \in \mathrm{GL}(n,\mathbb{C}).$$

of $\mathrm{GL}(n,\mathbb{C})$ on the space of all pluriharmonic forms.

Hence one is lead to the question of decomposition of this representation into irreducible constituents. This proplem has been solved completely by KASHIWARA-VERGNE [KV]. In their approach it is essential to do more, namely to decompose the space of pluriharmonic polynomials under the action (6.3) of the bigger group $O(r,\mathbb{C}) \times \mathrm{GL}(n,\mathbb{C})$.

In the appendix to this section, we describe some of their results. We don't give any proofs, because these results are not really needed in the following.

We reformulate 5.6 in the case of a harmonic form.

6.7 Proposition. *Let*

$$P : \mathbb{C}^{(r,n)} \longrightarrow \mathcal{Z}$$

be a harmonic form with respect to the representation

$$\varrho_0 : \mathrm{GL}(n, \mathbb{C}) \longrightarrow \mathrm{GL}(\mathcal{Z}).$$

Then the (vector valued) theta series

$$\vartheta_P \begin{bmatrix} U \\ V \end{bmatrix}(S; Z) = \sum_{G \in \mathbb{Z}^{(r,n)}} P\big(S^{1/2}(G+U)\big) \exp \pi i \sigma \big(S[G+U]Z + 2V'(G+U)\big)$$

satisfies the transformation formula

$$\widetilde{\vartheta}_P \begin{bmatrix} UD' - S^{-1}VC' \\ -SUB' + VA' \end{bmatrix}(S, M\langle Z\rangle) = v_S(M)\varrho(CZ + D)\widetilde{\vartheta}_P \begin{bmatrix} U \\ V \end{bmatrix}(S; Z)$$

for all $M \in \Gamma_n(S)$.

We recall the notation (I sec.4)

$$\varrho(A) = \varrho_0(A) \det(A)^{r/2}.$$

Now we assume that besides S, the characteristic U, V is rational. It is obvious that there exists a "level" such that $\exp \pi i \sigma \big(S[G+U]Z + 2V'(G+U)\big)$ will not change, if one replaces

$$U \longmapsto U + qX, \quad V \longmapsto V + qY, \qquad X, Y \text{ integral.}$$

An immediate consequence is

6.8 Corollary. *Let S, U, V be rational. Then there exists a congruence subgroup Γ such that the series*

$$\vartheta_P \begin{bmatrix} U \\ V \end{bmatrix}(S; Z)$$

is a modular form for each harmonic form P. More precisely:

$$\vartheta_P \begin{bmatrix} U \\ V \end{bmatrix}(S; Z) \in [\Gamma, \varrho, v_S].$$

Of course the same method gives precise information about possible levels q. We collect some results:

6.9 Corollary. *Assume that S is even (that means integral with even diagonal) and that q is a natural number such that qS^{-1} is even too. Then the function*

$$f(Z) = \vartheta_P \begin{bmatrix} 0 \\ 0 \end{bmatrix}(S; Z)$$

satisfies

$$f(M\langle Z\rangle) = v_S(M)\varrho(CZ + D)f(Z)$$

for all $M \in \Gamma_{n,0}[q]$ (I sec1).

The proof follows from 4.5 and 6.7. A similar result is

6.10 Corollary. *Assume that S is integral and that q is a natural number such that qS^{-1} is integral too. Then the function*

$$f(Z) = \vartheta_P \begin{bmatrix} 0 \\ 0 \end{bmatrix} (S; Z)$$

satisfies

$$f(M\langle Z \rangle) = v_S(M)\varrho(CZ + D)f(Z)$$

for all $M \in \Gamma_{n,0,\vartheta}[q]$ (I sec1).

The following variant will prove to be very important in our theory of singular modular forms.

6.11 Corollary. *Assume that S is integral and that q is a natural number such that $q^2 S^{-1}$ is integral too. Let furthermore $V = V^{(r,n)}$ be an integral matrix such that*

$$S^{-1}[V + qX]$$

is integral for all integral X. Then the function

$$f(Z) = \sum_{G \in \mathbb{Z}^{(r,n)}} P(S^{1/2}G) \exp \frac{\pi i}{q} \sigma \big(S[G]Z + 2V'G \big)$$

satisfies

$$f(M\langle Z \rangle) = v_{S/q}(M)\varrho(CZ + D)f(Z)$$

for all $M \in \Gamma_n[q, 2q]$ (I sec.1).

Proof. We know (I sec.1)

$$\Gamma_n(S/q) \subset \Gamma_n[q, 2q].$$

Because of 6.7 we only have to show

$$\vartheta_P \begin{bmatrix} 0 \\ V \end{bmatrix} (S; Z) = \vartheta_P \begin{bmatrix} 2S^{-1}VC \\ VD/q \end{bmatrix} (S; Z).$$

In the exponent of the general term of the second series occurs

$$S[G + S^{-1}VC].$$

Transformation of the summation variable

$$G \longmapsto G - S^{-1}VC$$

gives the desired identity, if one makes use of the fact that

$$(D - E)/q \quad \text{and} \quad S^{-1}[V]$$

are integral.

One might feel that our proof of the transformation formalism is too complicated. In most cases one is only interested in harmonic forms and not arbitrary polynomial coefficients. In that case easier proofs might be possible. In some special cases this has been worked out. A proof of the result 6.7 for example can be found in [Mu]. For special harmonic forms, 6.9 has been proved in [AM]. But for our purposes it is necessary to prove converse theorems: Under certain assumptions we will see:

If a theta series with a polynomial coefficient represents a modular form, then the coefficient is automatically harmonic.

For such converse theorems our more general approach seems at least partly unavoidable.

Appendix to sec.6
The theory of harmonic forms

We describe some results of KASHIWARA-VERGNE [KV] and some consequences of them. We do not really need them in what follows. Nevertheless they are very useful to get a complete picture.

We denote by

$$\mathcal{H}(r, n, k)$$

the space of pluriharmonic polynomials

$$P : \mathbb{C}^{(r,n)} \longrightarrow \mathbb{C},$$

which are homogenous of degree k. This space is finite dimensional. The group $GL(n, \mathbb{C})$ acts on this space by

$$(A \circ P)(X) = P(XA) \quad (A \in GL(n, \mathbb{C}); \quad P \in \mathcal{H}(r, n, k)).$$

As we already mentioned, the description of all harmonic forms is equivalent with the decomposition of $\mathcal{H}(r, n, k)$ into irreducible components. We also mentioned that it is important to include the action of the orthogonal group $O(r, \mathbb{C})$, which acts by

$$(B \circ P)(X) = P(B^{-1}X).$$

This means that we have an action of

$$O(r, \mathbb{C}) \times GL(n, \mathbb{C})$$

on $\mathcal{H}(r, n, k)$ defined by

$$((A, B) \circ P)(X) = P(A^{-1}XB).$$

It is advantageous first to consider the action of $O(r, \mathbb{C})$ only. This is a reductive algebraic group. The rational representations of this orthogonal group can be described by a highest weight theory, similar to and explained in some detail for $GL(n, \mathbb{C})$ (s. chap. I, appendix to sec.6). We don't give the details for the orthogonal group here.

We denote by

$$O(r, \mathbb{C})^{\wedge}$$

the set of isomorphism classes of irreducible rational representations of $O(r, \mathbb{C})$. (This set is in one-to-one correspondence with the unitary dual of the real orthogonal group.)

We denote by

$$\Sigma(n, k) \subset O(r, \mathbb{C})^{\wedge}$$

the subset of all representations, which occur in $\mathcal{H}(r, n, k)$.

In the paper [KV] the set

$$\Sigma(n) = \bigcup_{k} \Sigma(n, k)$$

is characterized by several equivalent properties. The most important is the description through the list of the highest weights.

For each $\lambda \in \Sigma(n, k)$ one considers the isotypic component

$$\mathcal{H}(\lambda) \subset \mathcal{H}(r, n, k).$$

By definition it is the sum of all irreducible subspaces of type λ. The $O(r, \mathbb{C})$-module $\mathcal{H}(\lambda)$ is a finite direct sum of "copies" of λ.

It is trivial that the group $GL(n, \mathbb{C})$ acts on $\mathcal{H}(\lambda)$ through translation from the right.

6.12 Proposition. *The representation of*

$$O(r, \mathbb{C}) \times \mathrm{GL}(n, \mathbb{C})$$

on the λ-isotypic component

$$\mathcal{H}(\lambda) \subset \mathcal{H}(r, n, k) \quad \left(\lambda \in \Sigma(n, k)\right)$$

is irreducible.

The proof of this fundamental result is tricky but simple and short ([KV], (5.7)).

The proposition can also be formulated as follows:

The $O(r, \mathbb{C}) \times \mathrm{GL}(n, \mathbb{C})$-module $\mathcal{H}(\lambda)$ is the tensor product of λ and an irreducible representation of $\mathrm{GL}(n, \mathbb{C})$.

Let $\mathrm{GL}(n, \mathbb{C})^{\wedge}$ denote the set of isomorphy classes of irreducible rational representations of $\mathrm{GL}(n, \mathbb{C})$. We reformulate 6.12 as follows:

6.13 Corollary. *There exists a map*

$$\Sigma(n, r) \longrightarrow \mathrm{GL}(n, \mathbb{C})^{\wedge},$$
$$\lambda \longmapsto \tau(\lambda),$$

such that the λ-isotypic component $\mathcal{H}(\lambda)$ is isomorphic to

$$\lambda \otimes \tau(\lambda)$$

as $O(r, \mathbb{C}) \times \mathrm{GL}(n, \mathbb{C})$-module.

6.14 Corollary.

$$\mathcal{H}(r, n, k) = \bigoplus_{\lambda \in \Sigma(n,k)} \lambda \otimes \tau(\lambda).$$

In a next step KASHIWARA and VERGNE write down a complete list of highest weight vectors in $\mathcal{H}(r, n, k)$ for the action of $O(r, \mathbb{C}) \times \mathrm{GL}(n, \mathbb{C})$. From this list they deduce the description of the set $\Sigma(n, k)$ and of the map

$$\lambda \longmapsto \tau(\lambda).$$

As a central consequence of those calculations they obtain:

6.15 Proposition. *The map*

$$\lambda \longmapsto \tau(\lambda)$$

is injective.

6.16 Corollary. *The isotypic components of $\mathcal{H}(r, n, k)$ with respect to the action of $O(r, \mathbb{C})$ (from the left) and with respect to the action of $\mathrm{GL}(n, \mathbb{C})$ (from the right) are the same!*

We see that the decomposition of $\mathcal{H}(r, n, k)$ under $O(r, \mathbb{C})$ and $GL(n, \mathbb{C})$ are tied closely together. In proposition 6.12 the roles between the two groups can be interchanged. We formulate this in a slightly different manner.

Let

$$\varrho_0 : GL(n, \mathbb{C}) \longrightarrow GL(\mathcal{Z})$$

be an irreducible rational representation. We denote by

$$\mathcal{H}(r, \varrho_0)$$

the space of all harmonic forms

$$P : \mathbb{C}^{(r,n)} \longrightarrow \mathcal{Z}$$

in the sense of 6.5. This $GL(n, \mathbb{C})$-module is obviously isomorphic with the ϱ_0-isotypic component of $GL(n, \mathbb{C})$ in $\mathcal{H}(r, n)$. From 6.16 follows:

6.17 Proposition. *The space $\mathcal{H}(r, \varrho_0)$ is irreducible under the action*

$$P(X) \longrightarrow P(B^{-1}X); \quad B \in O(r, \mathbb{C}),$$

of the orthogonal group.

Corollary. *If P is a non vanishing harmonic form in $\mathcal{H}(r, \varrho_0)$, then the whole space is spanned by the forms*

$$X \longmapsto P(BX); \quad B \in O(r, \mathbb{C}).$$

Examples of harmonic forms

We assume now that ϱ_0 is polynomial. For each matrix $A \in \mathbb{C}^{(r,n)}$ and for each vector $a \in \mathcal{Z}$ the function

$$P(X) := \varrho_0(X'A)a$$

is a form, i.e.

$$P(XB) = \varrho_0(B')P(X) \quad \text{for all } B \in GL(n, \mathbb{C}).$$

The question is whether it is a harmonic form.

6.18 Remark. *Assume that $A \in \mathbb{C}^{(r,n)}$ is a matrix with the property*

$$A'A = 0.$$

Then for each vector $a \in \mathcal{Z}$ the polynomial

$$P(X) := \varrho_0(X'A)a$$

is a harmonic form with respect to ϱ_0.

Proof. For any C^∞-function

$$f : \mathbb{C}^{(r,n)} \longrightarrow \mathbb{C},$$

the chain rule yields

$$\sum_{\nu=1}^{r} \frac{\partial}{\partial x_{\nu i}} \frac{\partial}{\partial x_{\nu k}} f(X'A)$$

$$= \sum_{\alpha=1}^{n} \sum_{\beta=1}^{n} \frac{\partial^2 f}{\partial u_{i\alpha} \partial u_{i\beta}} \Big|_{U=X'A} \cdot \sum_{\nu=1}^{n} a_{\nu i} a_{\nu k}.$$

This will be 0 if $A'A = 0$.

If we assume $r \geq 2n$, a matrix $A \in \mathbb{C}^{(r,n)}$ with the properties

$$A'A = 0; \quad \text{rank } A = n$$

exists, for example

$$A = \begin{pmatrix} E^{(n)} \\ i E^{(n)} \\ 0 \end{pmatrix}.$$

For such an A, there exists a matrix $X \in \mathbb{C}^{(r,n)}$ with $A'X \in \mathrm{GL}(n, \mathbb{C})$. The orthogonal group acts on the matrices A by multiplication from the left. From 6.17 we obtain:

6.19 Proposition. *Assume*

$$r \geq 2n.$$

The space $\mathcal{H}(r, \varrho_0)$ of harmonic forms with respect to an irreducible polynomial representation ϱ_0 is generated by the special polynomials

$$X \longmapsto \varrho_0(X'A)a, \quad A'A = 0.$$

In the case of the one dimensional representation $\det^k$, this result has been proved independently of the theory of KASHIWARA-VERGNE by MAASS [Ma2]. He also observed that in the case $k = 1$ the condition $A'A = 0$ is not necessary for harmonicity. He also had to restrict to $r \geq 2n$. But as WEISSAUER pointed out, the theory of KASHIWARA-VERGNE gives more, namely, the restriction $r \geq 2n$ is superfluous in the one dimensional case:

6.20 Proposition. *The space of harmonic forms*

$$P : \mathbb{C}^{(r,n)} \longrightarrow \mathbb{C}$$

with the properties

a) $$P(XA) = (\det A)^k P(X),$$
b) $$\Delta P = 0$$

is generated by the polynomials

$$\det(B'X)^k,$$

where B belongs to $\mathbb{C}^{(r,n)}$ with

$$B'B = 0 \quad \text{if} \quad k > 1.$$

7   Some multiplier systems

The following computation of special Gauss sums is due to ANDRIANOV and MALOLETKIN [AM]. More precisely, they reduce it to the classical case $n = 1$ which can be found for example in EICHLER's book [Ei].

Let S be a symmetric rational $r \times r$−matrix The Gauss sum which we have to consider is

$$v_S(M) = \sum_{G \bmod D} \exp \pi i \sigma(BD^{-1}S[G]).$$

It is well defined for example if

$$M = \begin{pmatrix} A & B \\ C & D \end{pmatrix}$$

is contained in the group $\Gamma_n(S)$ (s. sec.4). Then it is a special value of the theta multiplier system

$$v_S(M) = v_\vartheta(M^S).$$

In the following q denotes a natural number (the "level").

7.1 Proposition. *Assume that the number r of "variables" of S is even. Assume furthermore that either*

1) S and qS^{-1} are even and

$$M \in \Gamma_{n,0}[q] \quad (s.\ I.\ sec.1)$$

or

2) S and qS^{-1} are integral, q is even and

$$M \in \Gamma_{n,0,\vartheta}[q] \quad (s.\ I.\ sec.1).$$

Then

$$v_S(M) = \operatorname{sgn}(\det D)^{r/2} \left(\frac{(-1)^{r/2} \cdot \det S}{|\det D|} \right),$$

where $\left(\frac{\cdot}{\cdot} \right)$ denotes the generalized Legendre symbol (s. I sec.5).

The first case can be found in the above mentioned paper of ANDRIANOV-MALOLETKIN [AM]. The variant 2) can be proved along the same lines [En]. The strategy of the proof is to reduce it to the case $n = 1$, which has been treated classically (for example s. [Ei]). The reduction rests on

7.2 Proposition. *The group $\Gamma_{n,0}[q]$ is generated by the following matrices:*

a) $\qquad \begin{pmatrix} U' & 0 \\ 0 & U^{-1} \end{pmatrix}; \quad U \in \mathrm{SL}(n, \mathbb{Z}),$

b) $\qquad \begin{pmatrix} E & H \\ 0 & E \end{pmatrix}; \quad H/q$ *integral,*

c) $\qquad \begin{pmatrix} E & 0 \\ H & E \end{pmatrix}; \quad H/q$ *integral,*

d) $\qquad \begin{pmatrix} E & 0 & 0 & 0 \\ 0 & a & 0 & b \\ 0 & 0 & 1 & 0 \\ 0 & c & 0 & d \end{pmatrix}; \quad \begin{pmatrix} a & b \\ c & d \end{pmatrix} \in \Gamma_{1,0}[q].$

7.3 Proposition. *Asume that q is even.*

The group $\Gamma_{n,0,\vartheta}[q]$ is generated by the following matrices:

a) $\qquad\qquad\qquad \begin{pmatrix} U' & 0 \\ 0 & U^{-1} \end{pmatrix};\quad U \in \mathrm{SL}(n,\mathbb{Z}),$

b) $\qquad\qquad\qquad \begin{pmatrix} E & H \\ 0 & E \end{pmatrix};\quad H/q \text{ even},$

c) $\qquad\qquad\qquad \begin{pmatrix} E & 0 \\ H & E \end{pmatrix};\quad H/q \text{ even},$

d) $\qquad\qquad\qquad \begin{pmatrix} E & 0 & 0 & 0 \\ 0 & a & 0 & b \\ 0 & 0 & 1 & 0 \\ 0 & c & 0 & d \end{pmatrix};\quad \begin{pmatrix} a & b \\ c & d \end{pmatrix} \in \Gamma_{1,0,\vartheta}[q].$

Proof (of 7.2 and 7.3). We will show that a given matrix M in the considered group can be transformed by multiplication from left and right with matrices of the type a), b), c) in finitely many steps into the form d). Because we may argue by induction, it is sufficient to bring M into the form

$$M = \begin{pmatrix} 1 & 0 & 0 & 0 \\ 0 & A_1 & 0 & B_1 \\ 0 & 0 & 1 & 0 \\ 0 & C_1 & 0 & D_1 \end{pmatrix}.$$

Multiplying a given matrix M in the considered group from the left and right with matrices of type a) , we can replace

$$D \longmapsto UDV,\quad U,V \in \mathrm{GL}(n,\mathbb{Z}).$$

Hence we can assume that D is a diagonal matrix with diagonal $d_1,\ldots d_n$. Let

$$c := \gcd(c_{11},\ldots c_{1n})$$

be the greatest common divisor of the first row of C, which of course is coprime to d_1. We write it as an integral linear combination

$$c = f_1 c_{11} + t_2 c_{12} + \ldots f_n c_{1n}.$$

With a number α, which will be choosen later, we consider

$$t_1 = \alpha f_1,\ldots t_n = \alpha f_n,$$

and put it into the second row and column of a (symmetric) matrix T, which conists of zeros elsewhere.

$$T = \begin{pmatrix} 0 & t_1 & 0 & \dots & 0 \\ t_1 & t_2 & t_3 & \dots & t_n \\ 0 & t_3 & 0 & \dots & 0 \\ \vdots & \vdots & \vdots & & \vdots \\ 0 & t_n & 0 & \dots & 0 \end{pmatrix}.$$

Now we replace

$$M \longmapsto M \cdot \begin{pmatrix} E & T \\ 0 & E \end{pmatrix} = \begin{pmatrix} * & * \\ C & \bar{D} \end{pmatrix}.$$

The first two elements in the first row of $\bar{D} = C + DT$ are

$$\bar{d}_{11} = d_1 + \alpha c_{12}, \quad \bar{d}_{12} = \alpha c.$$

So $\bar{D}$ is no longer a diagonal matrix. In the case 7.2, we choose $\alpha = q$. From $C \equiv 0 \bmod q$, we obtain that q and d_1 are coprime, hence

$$\gcd(\bar{d}_{11}, \bar{d}_{12}) = 1.$$

Therefore the first elementary divisor of $\bar{D}$ is 1. Using matrices of the type a) again, we can assume $d_1 = 1$. If we multiply the matrix M now from the right with a matrix

$$\begin{pmatrix} E & 0 \\ F & E \end{pmatrix}, \quad F/q \text{ integral,}$$

we may enforce that the first row of the (new) matrix C is zero. The same follows also for the first column from the symplectic relations. A similar argument works for B and the rest from the symplectic relations.

The case 7.3 is a little bit more involved. Here on has to take for α a suitable positive power of $2q$. If one makes use of the fact that q is even, an argument similar to the one in case 7.2 goes through.

Proof of 7.1. Because the number r of variables is even, the statement is an identity between characters and hence has to be proved only for the generators. The propositions 7.2, 7.3 give a reduction to the case $n = 1$, i.e to the calculation of the classical Gauss sum

$$\sum_{g \bmod d} \exp \pi i \frac{b}{d} S[g].$$

We now assume that q, hence d is odd. (The case of an even q is a little more complicated. Because we will not have to make use of this case we refer to [Ei] for details.) In the odd case one can use the following well-known lemma.

7.4 Lemma. *Let S be an integral symmetric matrix and d an odd integer. There exists a unimodular matrix $U \in \mathrm{SL}(n, \mathbb{Z})$, such that $S[U]$ is a diagonal matrix mod d.*

If $s_1, \ldots, s_n$ are the elements of such a diagonalization, we obtain an expression by the standard Gauss sums, which we already introduced in the appendix of I sec.5:

$$\sum_{g \bmod d} \exp \pi i \frac{b}{d} S[g] = \prod_{j=1}^{r} \sum_{x \bmod d} \exp \pi i \frac{b}{d} s_j x^2.$$

A simple application of our calculations of the multiplier systems is

7.5 Proposition. *Let $S \in \mathbb{Q}^{(r,r)}$ be a positive rational matrix. Assume that r is even. Then there exists a congruence subgroup Γ, such that v_S is trivial on Γ.*

7.6 Corollary. *Let $S \in \mathbb{Q}^{(r,r)}$ be a positive rational matrix. There exists a congruence subgroup such that v_S coincides on Γ with the rth power of the theta multiplier system.*

The group of all projectively rational symplectic matrices acts on the space of theta series of a given type as for example on

$$\sum \exp \pi i \big(Z[g + a] + 2b'(g + a) \big),$$

a, b rational. This follows for example from the fact that the group of projectively rational symplectic matrices is generated by the substitutions

$$Z \longmapsto -Z^{-1}; \quad Z \longmapsto Z + S; \quad Z \longmapsto tZ.$$

As a consequence of 7.6 we obtain that each conjugate of the theta multiplier system coincides with the theta multiplier system on a suitable congruence subgroup. This result has been formulated already in I 3.10.

III   Singular weights

1   Jacobi forms

Let $\mathcal{L}$ be a rational lattice of symmetric $n \times n$-matrices and let

$$f(Z) = \sum_{T \in \mathcal{L}} a(T) \exp \pi i \sigma(TZ)$$

be a Fourier series which converges on the upper half space $\mathbb{H}_n$. We decompose the variable Z into 4 blocs:

$$Z = \begin{pmatrix} Z_0 & Z_1 \\ Z_1' & Z_2 \end{pmatrix};$$

$$Z_0 \in \mathbb{H}_m, \ Z_1 \in \mathbb{C}^{(m,n-m)}, \ Z_2 \in \mathbb{H}_{n-m}.$$

Here m denotes a natural number

$$0 < m < n.$$

The simple observation

$$\sigma(TZ) = \sigma(T_0 Z_0) + 2\sigma(T_1' Z_1) + \sigma(T_2 Z_2)$$

allows us to rearrange the Fourier series in the following way:

$$f(Z) = \sum_{T_2} \varphi_{T_2}(Z_0, Z_1) \exp \pi i \sigma(T_2 Z_2),$$

where

$$\varphi_{T_2}(Z_0, Z_1) = \sum_{T = \begin{pmatrix} T_0 & T_1 \\ T_1' & T_2 \end{pmatrix}} a(T) \exp \pi i \sigma(T_0 Z_0 + 2 T_1' Z_1).$$

This series is the Fourier coefficient of $f(Z)$ with respect to the variable Z_2. Hence the series φ_{T_2} converges for all pairs (Z_0, Z_1), which are part of a matrix $Z \in \mathbb{H}_n$, i.e. for all

$$Z_0 \in \mathbb{H}_m; \quad Z_1 \in \mathbb{C}^{(m,n-m)}.$$

They are holomorphic functions

$$\varphi_{T_2} : \mathbb{H}_m \times \mathbb{C}^{(m,n-m)} \longrightarrow \mathbb{C}.$$

The functions φ_{T_2} are called the **Fourier-Jacobi coefficients** of f and the whole series is the Fourier-Jacobi expansion of f.

If f is a modular form, the Fourier-Jacobi coefficients inherit several transformation equations from f. These equations lead to the notion of a **Jacobi form**.

Before we can give the definition of a Jacobi form, we need certain ingredients:

1.1 Notation. *The group* $G(m,n)$
$$G(m,n) \subset \mathrm{GL}(n,\mathbb{C}) \quad (0 < m < n)$$
is the subgroup of $\mathrm{GL}(n,\mathbb{C})$ *consisting of all*
$$\begin{pmatrix} A & B \\ 0 & E \end{pmatrix} \in \mathrm{GL}(n,\mathbb{C}) \quad (A \in \mathrm{GL}(m,\mathbb{C})).$$

Obviously $G(m,n)$ is a semidirect product of $\mathrm{GL}(m,\mathbb{C})$ and the abelian group $\mathbb{C}^{(m,n-m)}$.

1.2 Assumption. *Let the following data be given:*

1) *Natural numbers*
$$0 < m < n; \quad q \equiv 0 \bmod 4.$$

2) *A symmetric semipositive integral matrix*
$$J \in \mathbb{Z}^{(n-m,n-m)}; \quad J \geq 0.$$

3) a) *An integer*
$$r \in \mathbb{Z};$$

   b) *a (rational) representation*
$$\varrho_0 : G(m,n) \longrightarrow \mathrm{GL}(\mathcal{Z}); \quad \dim_{\mathbb{C}} \mathcal{Z} < \infty.$$

We are not really interested in the pair (ϱ_0, r) but only in the equivalence class $[\varrho_0, r]$ with respect to the following equivalence relation (compare I4.2):
$$(\varrho_0, r) \equiv (\varrho'_0, r')$$
$$\Longleftrightarrow$$
$$\text{a)} \quad r \equiv r' \bmod 2,$$
$$\text{b)} \quad \varrho_0 \cdot \det^{(r-r')/2} = \varrho'_0.$$

If r is even, the equivalence class corresponds uniquely to the representation
$$\varrho = \varrho_0 \cdot \det^{r/2}.$$

In any case, the function
$$\varrho \begin{pmatrix} CZ+D & 0 \\ 0 & E \end{pmatrix} = \det(CZ+D)^{r/2} \varrho_0 \begin{pmatrix} CZ+D & 0 \\ 0 & E \end{pmatrix}$$

is well defined. In this connection it is good to make the following inductive convention about the choice of the two square roots:
$$\sqrt{\det(CZ+D)} = \sqrt{\det \left(\begin{pmatrix} C & 0 \\ 0 & 0 \end{pmatrix} \begin{pmatrix} Z & 0 \\ 0 & iE \end{pmatrix} + \begin{pmatrix} D & 0 \\ 0 & E \end{pmatrix} \right)}.$$

If D is invertible this is compatible with the choices made in the appendix to I sec.5, which we also use in the following. Especially
$$\varrho \begin{pmatrix} E & G \\ 0 & E \end{pmatrix} = \varrho_0 \begin{pmatrix} E & G \\ 0 & E \end{pmatrix}.$$

After these preparations we give the definition of a Jacobi form.

1.3 Definition. *A Jacobi form with respect to the given data (1.2)*

$$0 < m < n; \quad J \in \mathbb{Z}^{(n-m,n-m)}, \quad J \geq 0$$
$$\varrho = [\varrho_0, r] \quad (\varrho_0 : G(m,n) \longrightarrow \mathrm{GL}(\mathcal{Z}))$$

is a holomorphic function

$$\varphi : \mathbb{H}_m \times \mathbb{C}^{(m,n-m)} \longrightarrow \mathcal{Z},$$

which satisfies the following transformation formulae:

1) a) $\varphi(Z, W + G) = \varphi(Z, W),$

   b) $\varphi(Z, W + ZG) = \varrho \begin{pmatrix} E & -G \\ 0 & E \end{pmatrix} \exp(\dfrac{-\pi i}{q} \sigma\{J(2G'W + Z[G])\}) \varphi(Z, W),$

   if $G \equiv 0 \bmod q$ (in both cases)

2)   *For all*

$$M = \begin{pmatrix} A & B \\ C & D \end{pmatrix} \in \Gamma_m[q, 2q]$$

   one has

$$\varphi(M\langle Z\rangle, (CZ + D)'^{-1}W)$$
$$= \exp \frac{\pi i}{q} \sigma(JW'(CZ + D)^{-1}CW) v_\vartheta(M)^r \varrho \begin{pmatrix} CZ + D & CW \\ 0 & E \end{pmatrix} \varphi(Z, W).$$

The assumption $q \equiv 0 \bmod 4$ (s.1.2) implies that the multiplier $v_\vartheta(M)^r$ depends only on $r \bmod 2$. It is trivial, if r is even (I3.11). Of course such an assumption is not really necessary. But for our purposes, the question of levels will never be important. In a systematic theory of Jacobi forms such an assumption would be unnatural. We refer to [EZ], [Zi] for some steps into a general context.

The following remark may be considered as a first hint that the notion of a Jacobi form is natural.

1.4 Remark. *The symplectic group* $\mathrm{Sp}(m, \mathbb{R})$ *acts on the space*

$$\mathbb{H}_m \times \mathbb{C}^{(m,n-m)}$$

by

$$(Z, W) \longmapsto (M\langle Z\rangle, (CZ + D)'^{-1}W).$$

For each symmetric $(n - m) \times (n - m)$-matrix J the function

$$R(M, (Z, W)) = \exp \pi i \sigma \left(JW'(CZ + D)^{-1}CW \right)$$

satisfies the cocycle relation

$$R(MN, (Z, W)) = R(M, N(Z, W)) \cdot R(N, (Z, W)).$$

Proof. Let be

$$M = \begin{pmatrix} A & B \\ C & D \end{pmatrix}; \quad N = \begin{pmatrix} \tilde{A} & \tilde{B} \\ \tilde{C} & \tilde{D} \end{pmatrix}.$$

One has to prove the formula

$$\left((C\tilde{A} + D\tilde{C})Z + (C\tilde{B} + D\tilde{D})\right)^{-1}(C\tilde{A} + D\tilde{C})$$
$$= (\tilde{C}Z + \tilde{D})^{-1}\tilde{C} + (\tilde{C}Z + \tilde{D})^{-1}\left(C(\tilde{A}Z + \tilde{B})(\tilde{C}Z + \tilde{D})^{-1} + D\right)^{-1}C(\tilde{C}Z + \tilde{D})'^{-1}.$$

The proof can be given by a somewhat tedious calculation with the symplectic relations. Another type of argument runs as follows: It is sufficient to prove the formula for all M, N in some congruence subgroup. Therefore it suffices to find one non-trivial example of a Jacobi-form. Such forms will be obtained as Fourier-Jacobi coefficients of modular forms (1.7).

1.5 Lemma. *Assume that $\varphi(Z, W)$ is a Jacobi form. Then for each $M \in \mathrm{Sp}(n, \mathbb{Z})$ the function*

$$\tilde{\varphi}(Z, W) = R(M, (Z, W))^{-1} \varrho \begin{pmatrix} CZ + D & CW \\ 0 & E \end{pmatrix}^{-1} \varphi(M(Z, W))$$

is a Jacobi form too.

Proof. By trivial reasons the expression

$$\varrho_0 \begin{pmatrix} CZ + D & CW \\ 0 & E \end{pmatrix}$$

satisfies the cocycle relation. Hence condition 2) in 1.3 is an immediate consequence of 1.4. We have to prove 1)a)b). Because of 1.4 it is sufficient to give the proof for the generators of the modular group. The case of translations being trivial, we may restrict to the involution $M = I$. Now the conditions a) for $\tilde{\varphi}$ follows from the condition b) for φ and conversely. We omit the easy calculation.

Each Jacobi form is periodic with respect to both variables. Hence it admits a Fourier expansion

$$\varphi(Z, W) = \sum_{T=T'} \sum_R a(T, R) \exp \pi i \sigma(TZ + 2R'W)$$

1.6 Definition. *The Jacobi form is called regular at infinity, if*

$$a(T, R) \neq 0 \Longrightarrow \begin{pmatrix} T & R \\ R' & J \end{pmatrix} \geq 0.$$

It is called regular at all cusps, if for each $M \in \mathrm{Sp}(n, \mathbb{Z})$ the transformed $\tilde{\varphi}$ (s.1.5) is regular at infinity.

(J is the given matrix in the assumption 1.2. It is sometimes called the **index** of the Jacobi form.)

The definition of a Jacobi form looks so strange and artificial that we discuss now the main example which we introduced at the beginning of this section.

1.7 Lemma. *Let*

$$f : \mathbb{H}_n \longrightarrow \mathcal{Z}$$

be a modular form

$$f \in [\Gamma[q, 2q], \varrho, v_\vartheta^r].$$

We consider the Fourier-Jacobi expansion

$$f(Z) = \sum_{T_2} \varphi_{T_2}(Z_0, Z_1) \exp \frac{\pi i}{q} \sigma(T_2 Z_2).$$

with respect to some choice m, $0 < m < n$. Then

$$\varphi_{T_2} : \mathbb{H}_m \times \mathbb{C}^{(m, n-m)} \longrightarrow \mathcal{Z}$$

is a Jacobi form with respect to the data

$$J = T_2 \qquad (= index)$$

and to

$$[\varrho_0 | G(m, n), r].$$

Such Jacobi forms are regular at all cusps.

Proof. 1a) The periodicity of f implies the periodicity of φ.

1b) We use the transformation formula

$$f(Z[U]) = \varrho(U^{-1}) f(Z)$$

for

$$U = \begin{pmatrix} E & G \\ 0 & E \end{pmatrix}, \quad G \equiv 0 \mod q.$$

Because of

$$\begin{pmatrix} Z_0 & Z_1 \\ Z_1' & Z_2 \end{pmatrix} \begin{bmatrix} E & G \\ 0 & E \end{bmatrix} = \begin{pmatrix} Z_0 & Z_0 G + Z_1 \\ * & Z_0[G] + Z_1' G + G' Z_1 + Z_2 \end{pmatrix}$$

we obtain

$$\varphi(Z_0, Z_0 G + Z_1) \exp \frac{\pi i}{q} \sigma \big(T_2(Z_0[G] + 2G' Z_1) \big) = \varrho \begin{pmatrix} E & -G \\ 0 & E \end{pmatrix} \varphi(Z_0, Z_1).$$

Now we make use of the transformation property of the modular form f under substitutions of the type

$$\widetilde{M} = \begin{pmatrix} A & 0 & B & 0 \\ 0 & E & 0 & 0 \\ C & 0 & D & 0 \\ 0 & 0 & 0 & E \end{pmatrix}; \quad M = \begin{pmatrix} A & B \\ C & D \end{pmatrix}.$$

A simple calculation yields

$$\widetilde{M}\langle Z \rangle = \begin{pmatrix} M\langle Z_0 \rangle & (CZ_0 + D)'^{-1} Z_1 \\ * & Z_2 - Z_1'(CZ_0 + D_0)^{-1} C Z_1 \end{pmatrix}.$$

Now the desired formula is evident.

2 Jacobi forms and theta series

We consider a Jacobi form

$$\varphi : \mathbb{H}_m \times \mathbb{C}^{(m,n-m)} \longrightarrow \mathcal{Z}$$

as defined in 1.3. The periodicity 1a) in the second variable allows a Fourier expansion

$$\varphi(Z,W) = \sum_{H \in \mathbb{Z}^{(m,n-m)}} b_H(Z) \exp \frac{2\pi i}{q} \sigma(H'W).$$

The coefficients $b_H(Z)$ are holomorphic functions on $\mathbb{H}_m$ with values in $\mathcal{Z}$. From the transformation property 1b) we immediately obtain

$$b_{H+qGJ}(Z) = \exp\bigl(\pi i \sigma(qJZ[G] + 2H'ZG)\bigr) \cdot \varrho \begin{pmatrix} E & qG \\ 0 & E \end{pmatrix} b_H(Z).$$

Hence the coefficients b_{H+qGJ} $(G \in \mathbb{Z}^{(m,n-m)})$ are determined by b_H.

In the following, we make the restriction that J is invertible $(J > 0)$. We let H run through a system of representatives $\bmod\, qJ$, i.e through a system of representatives of the orbits

$$\{H + qGJ; \quad G \text{ integral}\}.$$

This system is finite. We obtain (under the assumption $J > 0$)

$$\varphi(Z,W) = \sum_{H \in \mathbb{Z}^{(m,n-m)}/q\mathbb{Z}^{(m,n-m)}J} \varphi_H(Z,W),$$

where

$$\varphi_H(Z,W) =$$
$$\sum_{G \text{ integral}} \varrho \begin{pmatrix} E & qG \\ 0 & E \end{pmatrix} b_H(Z) \exp \pi i \sigma\bigl(qJZ[G] + 2H'ZG/q + 2(H/q + GJ)'W\bigr).$$

This is a theta series with a general coefficient function

$$\varrho \begin{pmatrix} E & qG \\ 0 & E \end{pmatrix} b_H(Z).$$

At the moment we don't have any information about harmonicity of the coefficients. There are two possible ways of deriving the transformation formulae for this theta series.

1) We can use the general transformation formalism for theta series with arbitrary coefficient functions as developed in II sec.5.

2) We can use the second transformation formula in the definition 1.3 of a Jacobi form.

We will derive important information by comparing both!

Using the formula

$$\sigma(qJZ[G] + 2H'ZG) = \sigma(JZ[G + HJ^{-1}/q] - q^{-1}J^{-1}Z[H])$$

and defining

$$c_H(Z) = \varrho \begin{pmatrix} E & -HJ^{-1} \\ 0 & E \end{pmatrix} b_H(Z) \cdot \exp\left(-\frac{\pi i}{q}\sigma(J^{-1}Z[H])\right)$$

we may rewrite the series φ_H in the form

$$\sum_G \varrho \begin{pmatrix} E & G + HJ^{-1}/q \\ 0 & E \end{pmatrix} c_H(Z) \exp \pi i \sigma\big(JZ[G + HJ^{-1}/q] + 2J(G + HJ^{-1}/q)'W\big).$$

Now we observe

$$\sigma(JZ[\widetilde{G}]) = \sigma(Z[\widetilde{G}']J)$$

and rewrite the above series as theta series with characteristics

$$U = J^{-1}H'/q; \quad V = JW'$$

and with the coefficient function

$$P_H(G, Z) = \varrho \begin{pmatrix} E & q^{1/2}G'J^{-1/2} \\ 0 & E \end{pmatrix} c_H(Z),$$

namely

$$\varphi_H(Z, W) = \vartheta_{P_H} \begin{bmatrix} U \\ V \end{bmatrix} (qJ; Z).$$

2.1 Lemma. *Let*

$$\varphi : \mathbb{H}_m \times \mathbb{C}^{(m,n-m)} \longrightarrow \mathcal{Z}$$

be a Jacobi form as defined in 1.3. Let H run through a system of representatives $\mathrm{mod}\, q$. For each H there exists a holomorphic function

$$c_H : \mathbb{H}_m \longrightarrow \mathcal{Z}$$

such that

$$\varphi(Z, W) = \sum_H \varphi_H(Z, W),$$

where

$$\varphi(Z, W) = \vartheta_{P_H} \begin{bmatrix} J^{-1}H'/q \\ JW' \end{bmatrix} (qJ; Z),$$

$$P_H(G, Z) = \varrho \begin{pmatrix} E & q^{1/2}G'J^{-1/2} \\ 0 & E \end{pmatrix} c_H(Z) \quad (now \ G \in \mathbb{C}^{(n-m,m)}).$$

Next we investigate the second transformation law in 1.3 of a Jacobi form. We obtain

$$\varphi(Z, W) = v_\vartheta(M)^{-r} \varrho \begin{pmatrix} CZ+D & CW \\ 0 & E \end{pmatrix}^{-1}$$
$$\exp\left(-\pi i \sigma (JW'(CZ+D)^{-1}CW)\right) \varphi\left(M\langle Z\rangle, (CZ+D)'^{-1}W\right)$$

or

$$\sum_H \vartheta_{P_H} \begin{bmatrix} J^{-1}H'/q \\ JW' \end{bmatrix} (J; Z)$$

$$= v_\vartheta(M)^{-r} \varrho \begin{pmatrix} CZ+D & CW \\ 0 & E \end{pmatrix}^{-1} \exp\left(-\pi i \sigma\left(JW'(CZ+D)^{-1}CW\right)\right)$$

$$\cdot \sum_H \vartheta_{P_H} \begin{bmatrix} J^{-1}H'/q \\ JW'(CZ+D)^{-1} \end{bmatrix} (J; M\langle Z\rangle),$$

where M runs through a suitable congruence group Γ.

This formula is a theta transformation formula (the variable Z is replaced by $M\langle Z\rangle$). It is natural to compare it with the general transformation formula proved in II sec.6. This formula states

$$\vartheta_{P_H} \begin{bmatrix} U \\ V \end{bmatrix} (qJ; Z) =$$

$$v_J(M)^{-1} \cdot \exp \pi i \sigma(U'V - \tilde{U}'\tilde{V}) \cdot \det(CZ+D)^{(m-n)/2} \vartheta_{P_H^M} \begin{bmatrix} \tilde{U} \\ \tilde{V} \end{bmatrix} (qJ, M\langle Z\rangle),$$

where

$$\tilde{U} = (J^{-1}H'D' - W'C')/q,$$
$$\tilde{V} = -H'B' + JW'A'.$$

A simple calculation gives

$$q\sigma(U'V - \tilde{U}'\tilde{V}) =$$
$$\sigma(HW' - CWH'B' - DHW'A') - \sigma(DHJ^{-1}H'B') + \sigma(CWJW'A').$$

In the first line after the equality sign we apply the symplectic relation $A'D - B'C = E$ and make use of the fact that the trace of a product is invariant under cyclic permutation. The second line may be cancelled (replacing Γ by a smaller group). By the same reasoning we may assume

$$v_J(M) = v_\vartheta(M)^{n-m}, \quad v_\vartheta(M)^2 = 1.$$

Hence we obtain

2.2 Remark. *We have*

$$\sum_H \exp \frac{\pi i}{q} \sigma(-2B'CWH' + CWJW'A') v_\vartheta(M)^{r+m-n}$$

$$\det(CZ + D)^{(r+m-n)/2}.$$

$$\vartheta_{P_H^M} \begin{bmatrix} J^{-1}H'D'/q - W'C'/q \\ -H'B' + JW'A' \end{bmatrix} (qJ; M\langle Z\rangle)$$

$$= \varrho_0 \begin{pmatrix} CZ + D & CW \\ 0 & E \end{pmatrix}^{-1} \exp \frac{\pi i}{q} \sigma(-JW'(CZ + D)^{-1}CW) \cdot$$

$$\sum_H \vartheta_{P_H} \begin{bmatrix} J^{-1}H'/q \\ JW'(CZ + D)^{-1} \end{bmatrix} (qJ; M\langle Z\rangle).$$

The "spirit" of this formula is that it gives a relation between the polynomials

$$P_H \quad \text{and} \quad P_H^M,$$

which finally implies information about $c_H(Z)$ and about ϱ.

It is tedious to work out a simple form for this relation.

It is useful to observe – after writing out the theta series in 2.2 – that on both sides an exponential factor occurs which contains W as homogenous polynomial of degree 2.

We will show that these factors are the same on both sides. We compute the factors on both sides seperately:

1)   Left hand side

We find two terms in the exponential $\exp \frac{\pi i}{q}$, namely

a) $\sigma(CWJW'A')$

and, writing down the theta series explicitly

b) $\sigma(J[W'C']M\langle Z\rangle - 2CWJW'A')$.

2)   Right hand side:

The only quadratic term which occurs is

$$\sigma(-JW'(CZ + D)^{-1}CW).$$

We want to show that the sum of the terms 1a) and 1b) equals the term in 2). During this calculation we make use of the fact the trace of a product is invariant under cyclic permutation $(\sigma(AB) = \sigma(BA))$.

Notation

$$K := WJW' = J[W'].$$

The claim is

$$\sigma(KA'C) + \sigma(CKC'M\langle Z\rangle) - 2\sigma(KA'C) = \sigma(-K(CZ+D)^{-1}C).$$

Next we replace

$$M\langle Z\rangle \longmapsto Z.$$

The cocycle relation I 3.1 gives

$$\left(CM\langle Z\rangle^{-1} + D\right)^{-1} = (-C'Z + A').$$

Now the claimed identity reads

$$\sigma(CKC'Z) - \sigma(KA'C) = \sigma\big(-K(-C'Z+A')C\big),$$

which is an obvious identity.

Our next goal is to get rid of the terms which contain products of components of W with components of Z in 2.2

To do this, we begin with the trivial remark that in a theta series $\vartheta_P \begin{bmatrix} U \\ V \end{bmatrix} (S; Z)$ the characteristic U is determined only mod 1. Therefore we may replace

$$J^{-1}H'D' \longmapsto J^{-1}H'$$

in the characteristic of the left hand side in 2.2.

The identity 2.2 may now be rewritten as

$$\sum_H \exp\frac{\pi i}{q}\sigma(-2B'CWH')$$

$$\sum_{G\equiv J^{-1}H'D'/q \bmod 1} P_H^M\big((qJ)^{1/2}(G - W'C'/q), M\langle Z\rangle\big)$$

$$\exp\pi i\sigma(qJ[G]M\langle Z\rangle - 2CWJGM\langle Z\rangle + 2G'JW'A' + 2CWH'B'/q)$$

$$=$$

$$v_\vartheta(M)^{r+n-m}\varrho_0 \begin{pmatrix} CZ+D & CW \\ 0 & E \end{pmatrix}^{-1} \det(CZ+D)^{-(r+m-n)/2}$$

$$\sum_H \sum_{G\equiv J^{-1}H'D'/q} P_H(J^{1/2}G) \qquad \exp\pi i\sigma\big(qJ[G]M\langle Z\rangle + 2G'JW'(CZ+D)^{-1}\big).$$

Claim. We have

$$\sigma(-CWJGM\langle Z\rangle + 2G'W'A') = \sigma(G'JW'(CZ+D)^{-1}).$$

Proof. We replace

$$Z \longmapsto M^{-1}\langle Z \rangle.$$

Using the cocycle relation

$$\left(CM^{-1}\langle Z \rangle + D \right)^{-1} = (-C'Z + A'),$$

the claimed identity is obvious.

With the notation

$$\widetilde{W} = (CZ + D)W$$

we obtain

$$v_\vartheta(M)^{(r+m-n)} \varrho_0 \begin{pmatrix} CZ + D & CW \\ 0 & E \end{pmatrix} \det(CZ + D)^{(m-n)/2}.$$

$$\sum_{H} \sum_{G \equiv J^{-1}H'D'/q \bmod 1} P_H^M\left((qJ)^{1/2}(G - W'C'), M\langle Z \rangle\right)$$

$$\exp \pi i \sigma q J[G]M\langle Z \rangle \cdot \exp \pi i \sigma(G\widetilde{W})$$

$$=$$

$$\sum_{H} \sum_{G \equiv J^{-1}H'D'/q \bmod 1} P_H\left((qJ)^{1/2}G, M\langle Z \rangle\right)$$

$$\exp \pi i \sigma q J[G]M\langle Z \rangle \cdot \exp \pi i \sigma(G\widetilde{W}).$$

The right hand side of this equality is a Fourier series in the variable $\widetilde{W}$. But in the right hand side the "coefficients" seem to depend on W too. The following lemma shows that they do not:

2.3 Lemma. *Assume that*

$$a_g(w_1, \ldots w_n); \quad g \in \mathbb{Z}^n$$

is a system of polynomials of bounded degree such that the series

$$\sum a_g(w) \exp 2\pi i g' w$$

converges locally uniformly and is identically 0. Then all coefficients a_g vanish.

Proof. By an obvious induction argument it is sufficient to consider the case $n = 1$. After that one uses induction on the maximal degree of the polynomials $a_g(w)$ (assuming that not all are identically 0.) Since

$$\sum \left(a_g(w+1) - a_g(w) \right) \exp 2\pi i g w$$

is identically 0, the induction hypothesis implies $a_g(w+1) = a_g(w)$. But periodic polynomials are constant.

The above identity being an identity between Fourier series we may compare the Fourier coefficients:

2.4 Lemma. *The coefficient function*

$$P_H(G, Z) = \varrho \begin{pmatrix} E & q^{1/2} G' J^{-1/2} \\ 0 & E \end{pmatrix} c_H(Z)$$

and the transformed function

$$P_H^M, \quad M \text{ runs through some congruence group,}$$

are connected by the formula

$$P_H(G, M\langle Z \rangle)$$
$$= v_\vartheta(M)^{(r+m-n)} \varrho_0 \begin{pmatrix} CZ + D & 0 \\ 0 & E \end{pmatrix} \det(CZ + D)^{(r+m-n)/2} P_H^M(G, M\langle Z \rangle).$$

Up to now the calculations have been somewhat long but absolutely straightforward and trivial. But these calculations will have an important consequence, namely the polynomials $G \longmapsto P_H$ will turn out to be **harmonic**. For the proof of this fundamental fact we have to introduce a certain characteristic operator.

2.5 Definition. *For $Z \in \mathbb{H}_m$, $J \in \mathbb{R}^{(n-m,n-m)}$ such that J is symmetric and positive, and for a representation $\varrho : \mathrm{Sp}(n, \mathbb{R}) \longrightarrow \mathrm{GL}(\mathcal{Z})$, we define an operator $L(Z) \in \mathrm{End}(\mathcal{Z})$ by*

$$L(Z) = \int_{\mathbb{R}^{(m,n-m)}} \varrho \begin{pmatrix} E & q^{1/2} U J^{-1/2} \\ 0 & E \end{pmatrix} \exp\left(-\pi i U'(Z/i) U \right) dU.$$

2.6 Remark. *One has*

$$L(Z) = \varrho \begin{pmatrix} J(Z/i)^{1/2} & 0 \\ 0 & E \end{pmatrix} L\varrho \begin{pmatrix} J(Z/i)^{-1/2} & 0 \\ 0 & E \end{pmatrix},$$

and

$$L = L(iE)$$

is a unipotent operator (i.e. a strict upper triangular matrix with respect to a suitable basis).

We now want to investigate the coefficient function P_H (s. lemma 2.4). For sake of simplicity we omit the subscript H,

$$P = P_H; \quad c = c_H.$$

We assume for a moment that lemma 2.4 is valid for the involution $M = I$. From the definition of the action of I on the coefficient functions, we obtain

$$P^I(0, Z) = L(Z) c(-Z^{-1}).$$

Specializing the formula in lemma 2.4 to $G = 0$, one obtains

$$c(-Z^{-1}) = v_\vartheta(I)^{r+m-n} \varrho_0(-Z) \det(-Z)^{(r+n-m)/2} L(-Z^{-1}) c(Z).$$

Now we make use of the fact that I is involutive. This gives us that the function

$$\varrho_0(-Z^{-1}) L(Z) \varrho_0(Z) L(-Z^{-1}) c(Z)$$

is a constant multiple of $c(Z)$. From the definition, it is clear that the function $c(Z)$ is periodic with respect to some latice of (integral) symmetric matrices.

2.7 Lemma. *The function*

$$K(Z)c(Z), \quad K(Z) := \varrho_0(-Z^{-1})L(Z)\varrho_0(Z)L(-Z^{-1})$$

is a constant multiple of $c(Z)$. Therefore it is periodic with respect to a suitable lattice.

The above calculation seems only to give a proof of 2.7 under the assumption that lemma 2.4 is valid for $M = I$. For a general proof one has to make use of the fact that the full modular group – especially $M = I$ – acts on the space of Jacobi forms. We omit details.

From 2.7 and the periodicity of $c(Z)$, it follows that

$$K(Z + S)c(Z) = K(Z)c(Z),$$

where S runs through a certain lattice. But $K(Z)$ is a rational function in $(Z/i)^{1/2}$. It follows easily that such an identity holds for variable S, especially for $S = iE - Z$.

We obtain

$$K(Z)c(Z) = K(iE)c(Z) = L^2 c(Z) = C \cdot c(Z).$$

The operator L being unipotent (2.6) we obtain

$$Lc(Z) = c(Z).$$

2.8 Proposition. *The polynomials*

$$G \longmapsto P_H(G, Z)$$

are pluriharmonic.

Proof. The above calculation shows that these polynomials agree with their Gauss transform. From Schur's lemma it follows that their components with respect to a suitable basis of $\mathcal{Z}$ are homogenous. The claim follows from II 1.3, 3. corollary.

In II 6.4 the action of the symplectic group on pluriharmonic coefficient functions has been computed. As an application we obtain

The (pluriharmonic) coefficient function

$$P_H(G, Z) = \varrho \begin{pmatrix} E & q^{1/2} G' J^{-1/2} \\ 0 & E \end{pmatrix} c_H(Z)$$

satisfies the functional equation

$$P_H(G(CZ + D)', M\langle Z\rangle)$$
$$= v_\vartheta(M)^{r+n-m} \varrho_0 \begin{pmatrix} CZ + D & 0 \\ 0 & E \end{pmatrix} \det(CZ + D)^{(r+m-n)/2} P_H(G, Z)$$

for all M in a suitable congruence subgroup.

A simple reformulation of our result is

2.9 Theorem. *The coeficients $c(Z) := c_H(Z)$ have the following two properties:*

1) The polynomial

$$\varrho \begin{pmatrix} E & q^{1/2} G' J^{-1/2} \\ 0 & E \end{pmatrix} c(H)$$

is pluriharmonic in G.

2) $\quad c(M\langle Z \rangle) = v(M)^{r+n-m} = \varrho_0 \begin{pmatrix} CZ + D & 0 \\ 0 & E \end{pmatrix} \det(CZ + D)^{(r+m-n)/2} c(Z).$

It is very easy to prove a converse theorem, i.e.

Fourier Jacobi forms are nothing else but systems of certain usual modular forms.

A special case of this statement is a consequence of the so-called Shimura-isomorphism [Sh]. But in contrast to the Shimura-isomorphism we did not try to get hold of the precise levels.

3 Differential Operators

Besides the Fourier-Jacobi expansion, we need another tool for the classification of the singular weights, namely certain differential operators introduced by A. SELBERG and investgated by SELBERG and MAASS. For the proofs of the following results we refer to [Ma1].

We denote by

$$\mathcal{P}_n = \{ Y \in \mathbb{R}^{(n,n)}; \quad Y = Y' > 0 \}$$

the domain of all positive real symmetric $n \times n$-matrices and by $\mathcal{C}^\infty(\mathcal{P}_n)$ the space of infinitely differentiable functions on $\mathcal{P}_n$. We are interested in certain differential operators

$$M : \mathcal{C}^\infty(\mathcal{P}_n) \longrightarrow \mathcal{C}^\infty(\mathcal{P}_n).$$

The easiest ones are the partial derivatives $\partial/\partial y_{\mu\nu}$, which we collect in a symmetric matrix

$$\partial/\partial Y = (e_{\mu\nu}\partial/\partial y_{\mu\nu}),$$

$$e_{\mu\nu} = \frac{1}{2}(1 + \delta_{\mu\nu}) = \begin{cases} 1 & \text{if } \mu = \nu; \\ \frac{1}{2} & \text{else.} \end{cases}$$

Since the entries of this matrix are commuting operators, its determinant is well defined. The operator

$$M_n = \det(Y)\det(\partial/\partial Y) : \mathcal{C}^\infty(\mathcal{P}_n) \longrightarrow \mathcal{C}^\infty(\mathcal{P}_n)$$

has been investigated by SELBERG and by MAASS. We will need some of their results. For proofs we refer to [Ma1].

3.1 Lemma. *Let T be a symmetric $n \times n$-matrix,*

$$\det(\partial/\partial Y) \exp \sigma(TY) = (\det T) \exp \sigma(TY).$$

The proof is trivial. The transformation law of M_n under the substitutions $Y \longrightarrow Y^{-1}$ will be very important. This transformation induces an isomorphism

$$\sigma : C^\infty(\mathcal{P}_n) \longrightarrow C^\infty(Y_n),$$
$$f(Y) \longmapsto \big(Y \mapsto f(Y^{-1})\big).$$

We denote by

$$M_n^* = \sigma M_n \sigma$$

the transformed operator.

3.2 Proposition. *One has*

$$M_n^* = (-1)^n \det(Y)^{\frac{n-1}{2}} M_n \det(Y)^{-\frac{n-1}{2}}.$$

The last result about M_n we need is the construction of certain eigenforms of M_n.

3.3 Lemma. *Each symmetric positive real matrix Y admits a so-called Jacobi decomposition*

$$Y = T[B] = B'TB,$$

where B is a strict upper triangular matrix and T is a diagonal matrix with positive entries. This decomposition is unique; the "Jacobi coordinates" T, B are differentiable functions of Y.

For each system $\mathbf{r} = (r_1 \ldots r_n)$ of real numbers we may consider the function

$$h_{\mathbf{r}} : \mathcal{P}_n \longrightarrow \mathbb{R},$$
$$h_{\mathbf{r}}(Y) = \prod_{i=1}^{n} t_i^{r_i}.$$

3.4 Proposition. *The functions $h_{\mathbf{r}}(Y)$ are eigenforms of M_n:*

$$M_n h_{\mathbf{r}}(Y) = \prod_{i=1}^{n} \left(\frac{r_i + i - 1}{2}\right) h_{\mathbf{r}}(Y).$$

We give an application of this formula.

Let

$$\varrho : \mathrm{GL}(n, \mathbb{C}) \longrightarrow \mathrm{GL}(m, \mathbb{C})$$

be an irreducible representation with highest weight $\mathbf{r} = (r_1, \ldots r_n)$.

3.5 Lemma. *Let r be a real number such that*

$$M_n\Big(\det(Y)^r \varrho(Y)\Big) = 0.$$

Then

$$\prod_{i=1}^{n}\Big(r_i + r + \frac{1-i}{2}\Big) = 0.$$

Proof. Let v_0 be the highest weight vector for ϱ and $\langle \cdot, \cdot \rangle$ be a $\mathbb{C}$-bilinear form with $\langle v_0, v_0 \rangle \neq 0$. Then since

$$\langle v_0, \varrho(Y)v_0 \rangle = \langle v_0, v_0 \rangle h_r(Y)$$

we obtain

$$\langle v_0, v_0 \rangle \prod_{i=1}^{n}\Big(r_i + r + \frac{1-i}{2}\Big) = M_n \langle v_0, \det(Y)^r \varrho(Y)v_0 \rangle = 0.$$

Now lemma 3.4 is proved.

4 The classification of singular weights

We now have all the tools necessary to describe the singular weights. In this section

$$\varrho_0 : \mathrm{GL}(n, \mathbb{C}) \longrightarrow \mathrm{GL}(\mathcal{Z})$$

denotes an irreducible reduced representation (i.e. ϱ_0 is polynomial but $\varrho_0(A)/\det A$ is not polynomial). We consider modular forms both of integral and half integral weight $r/2$, $r \in \mathbb{Z}$, more precisely modular forms with respect to

$$\varrho(A) = \varrho_0(A)(\det A)^{r/2}.$$

4.1 Definition. *A modular form*

$$f(Z) = \sum a(T) \exp \pi i \sigma(TZ)$$

is called singular, if

$$a(T) \neq 0 \implies \det T = 0.$$

4.2 Definition. *Let f be a non-zero singular modular form. The rank of f is*

$$\mathrm{rank}\, f = \max\{\mathrm{rank}\, T; \quad a(T) \neq 0\}.$$

Of course
$$\text{rank } f < n.$$

The main result about singular weights is

4.3 Theorem. *A non-zero modular form with respect to ϱ is singular if and only if*

$$r < n.$$

A refinement of 4.3 (actually a corollary) states

4.4 Theorem. *Let f be a non-zero singular modular form. Then*

$$\text{rank } f = n - r.$$

1. Corollary. *If a non-zero modular form exists in $[\Gamma, \varrho, v]$, then all modular forms in this space are singular.*

2. Corollary. *Each conjugate $f|_\varrho M$ (M projectively rational) of a singular modular form is singular.*

Proof of 4.3. 1.step: We assume that f is a non-vanishing singular modular form. We show $r < n$.

A modular form is singular if and only if it is annihilated by the operator $L :=$ $\det(\partial/\partial Z)$. For all M in a suitable congruence subgroup we obtain $L\varrho(CZ + D)^{-1}f(M\langle Z\rangle) = 0$ or $L^M f(Z) = 0$, where L^M denotes a certain transformed operator. It is not neessary to work out an explicit formula for the transformed operator. What we need is that it depends polynomially on M. This follows easily from the explicit transformation formula for the involution I (s.3.2) in connection with the "Bruhat decomposition", which we used for example in the appendix of sec.5 in chap.I. The formula $L^M f(Z) = 0$ therefore is valid for all $M \in \text{Sp}(n, \mathbb{R})$. We apply this in the case $M = I$. The explicit transformation formula 3.3 for the operator $M_n = \det(Y)\det(\partial/\partial Y)$ gives us

$$M_n \det(Y)^{\frac{n-1}{2}} \varrho(Y)g(iY^{-1}) = 0,$$

where $g(Z) = \varrho(-Z^{-1})f(-Z^{-1})$ denotes the conjugate modular form. If we denote by $b(T)$ their Fourier coefficients, we obtain

$$M_n \det(Y)^{\frac{n-1}{2}} \varrho(Y)b(T) \exp \pi i\left(-TY\right) = 0.$$

The function $\det(Y)^{\frac{n-1}{2}}\varrho(Y)$ is homogenous in Y. From the nature of M_n easily follows

$$M_n \det(Y)^{\frac{n-1}{2}} \varrho(Y)b(T) = 0.$$

The vector space generated by all coefficients $b(T)$ is invariant under all $\varrho_0(U)$, where U belongs to a certain congruence subgroup. Therefore it is invariant under ϱ_0. We assumed that ϱ_0 is irreducible and that f does not vanish identically. We finally obtain the relation

$$M_n \det(Y)^{\frac{n-1}{2}} \varrho(Y) = 0.$$

The claim now follows from 3.5.

2.step: We assume $r < n$. We want to show that a modular form f is singular, equivalently

$$g(Z) := \det(\partial/\partial Z) f(Z) = 0.$$

The subspace of $\mathcal{Z}$, which is generated by the values of $g(Z)$, is invariant under $\varrho_0(U)$, where U runs through a congruence subgroup. As a consequence this subspace is invariant under ϱ_0. By assumption ϱ_0 is irreducible (especially $\mathcal{Z} \neq 0$). Therefore it is sufficient to construct a proper subspace $\mathcal{Z}' \subset \mathcal{Z}$, $\mathcal{Z}' \neq \mathcal{Z}$, such that the values of g are contained in $\mathcal{Z}'$.

Construction of $\mathcal{Z}'$. The representation

$$\mathrm{GL}(n-1,\mathbb{C}) \longrightarrow \mathcal{Z},$$

$$A \longmapsto \varrho_0 \begin{pmatrix} A & 0 \\ 0 & 1 \end{pmatrix}$$

is no longer irreducible. We decompose it into irreducible components. The subspace $\mathcal{Z}''$ is the sum of all components with heighest weight $(r_2, \dots r_n)$. The sum of the remaining components is $\mathcal{Z}'$. We have

$$\mathcal{Z} = \mathcal{Z}' \oplus \mathcal{Z}''.$$

We have to show that $\mathcal{Z}'' \neq 0$. For example the subspace of invariants of the group $\left\{ \begin{pmatrix} A & 0 \\ c' & 1 \end{pmatrix} \right\}$ is contained in $\mathcal{Z}''$. We decompose the modular form f,

$$f(Z) = f_1(Z) + f_2(Z); \quad f_1(Z) \in \mathcal{Z}', \ f_2(Z) \in \mathcal{Z}''.$$

Our claim is $g_2(Z) := \det(\partial/\partial Z) f_2(Z) = 0$. To show this we use the results about the Fourier Jacobi expansion, where the expansion is taken for the parameter $m = 1$. In sec.2 we investigated only Fourier Jacobi coefficients with respect to invertible indices J. But the other ones are killed by the operator $\det(\partial/\partial Z)$ because for semidefinite matrices one has

$$T = \begin{pmatrix} J & * \\ * & * \end{pmatrix}; \quad \det J = 0 \Rightarrow \det T = 0.$$

From the main result of sec.2 follows that the functions $c(Z)$ which occur in the expansion of the Fourier Jacobi coefficients of f_1 are modular forms of weight $r - 1 - n$. From our assumption we know $r - 1 - n \leq 0$. We know that such forms vanish identically, if the representation $[r_2, \dots r_n]$ is not trivial. So in this case the proof is already complete. In the trivial case $(r_2 = \dots = r_n = 0)$ the functions $c(Z)$ are constants. A glance to the Fourier Jacobi expansion of f_2 shows that in this case only non vanishing Fourier coefficients with respect to indices of the type

$$T = \begin{pmatrix} T_1 & g \\ g' & T_1[g] \end{pmatrix}$$

occur. Such indices have determinant 0 and are killed by $\det(\partial/\partial Z)$. The proof of theorem 4.3 is complete.

Theorem 4.4 is a corollary of theorem 4.3. One makes use of the Siegel Φ-operator. This operator has to be applied to all conjugates

$$f^U(Z) = \varrho(U')f(Z); \quad U \in \mathrm{GL}(n, \mathbb{Z}).$$

Assume that f is a singular modular form of rank $n - t$. Then we know that there exists a matrix $U \in \mathrm{GL}(n, \mathbb{Z})$, such that $f^U|\Phi^t$ does not vanish identically. We have two further informations:

a) $r_t = \ldots = r_n$ (I 6.4).

b) $f^U|\Phi^{t-1}$ is still singular.

c) $f^U|\Phi^t$ is not singular.

Now theorem 4.4 is an immediate consequence of theorem 4.3. The two corollaries are trivial.

In these notes we only considered weights r, such that $2r$ is integral. This is a natural condition for the theory of theta series but not necessary for the definition of a modular form. The expression $\det(CZ + D)^{r/2}$ can be defined holomorphically for all real r and the notion of a multiplier system $v(M)$ on a congruence subgroup too. One has to demand that $v(M)\det(CZ + D)^{r/2}$ satisfies the cocycle relation. Hereafter the definition of a modular form with respect to

$$\text{``}\varrho(A) = \varrho_0 \det(A)^{r/2}\text{''}$$

is obvious. We call $r/2$ the weight of the modular form, if ϱ_0 is reduced I 4.1. The proof of theorem 4.3 shows more, namely:

4.5 Theorem. *Let f be a non-vanishing modular form of weight $r/2$, $r < n$. Then r is integral.*

This theorem has an interesting geometric application. Let $\Gamma \subset \mathrm{Sp}(n, \mathbb{R})$ be any congruence subgroup. In the case $n \geq 3$, it is known that each closed analytic subvariety of codimension 1 is the precise divisor of a (scalar valued) modular form. This result follows from the investigation of the singular cohomology. In principle the result is due to GARLAND. His results have been improved by BOREL and WALLACH. From their theory the bound $n \geq 4$ can be derived. The improvement $n \geq 3$ is due to WEISSAUER (unpublished).

We call a (scalar valued) modular form a **prime form** if its divisor is irreducible (with multiplicity 1). We call f an **absolutely prime form** if it is a prime form with respect to arbitrary small congruence subgroups $\Gamma_0 \subset \Gamma$. The theory of singular modular forms together with the above mentioned geometric result implies the existence of absolutely prime forms:

4.6 Theorem. *Each non-vanishing scalar valued modular form of weight $1/2$ is an absolutely prime form.*

4.7 Theorem. *Let f be a non vanishing scalar valued modular form on Γ. There is up to the order a unique decomposition*

$$f = f_1 \cdot \ldots \cdot f_m,$$

where f_j $(1 \leq j \leq m)$ are absolutely prime forms on some congruence subgroup $\Gamma_0 \subset \Gamma$.

IV Singular modular forms and theta series

1 The big singular space

We fix three natural numbers

$$n \quad (= \text{degree}),$$
$$r \quad (= 2 \cdot \text{weight}),$$
$$q \quad (= \text{level})$$

and assume

$$r < n \quad (\text{singular case}).$$

Let furthermore

$$\varrho_0 : \text{GL}(n, \mathbb{C}) \longrightarrow \text{GL}(\mathcal{Z}), \quad \dim_{\mathbb{C}} \mathcal{Z} < \infty,$$

be an irreducible reduced representation on a finite dimensional complex vector space. ("Reduced" means that ϱ_0 is polynomial but does not vanish identically on the determinant surface "$\det A = 0$".)

We consider Fourier series converging on the Siegel upper half space of the following type:

$$f(Z) = \sum_{T=T' \text{ integral}} a(T) \exp \frac{\pi i}{q} \sigma(TZ),$$

where the coefficients $a(T)$ are in the vector space $\mathcal{Z}$. Hence the function f is vector valued,

$$f : \mathbb{H}_n \longrightarrow \mathcal{Z}.$$

1.1 Definition. *We denote by*

$$\mathbf{P} = \mathbf{P}(n, r, q, \varrho_0)$$

the space of all Fourier series converging on $\mathbb{H}_n$ and with the following properties:

a) $a(T) \neq 0 \Longrightarrow T \geq 0$ *(positive semidefinite),*

b) $a(T) \neq 0 \Longrightarrow \text{rank}\,(T) \leq r$,

c) $a(T[U]) = \varrho_0(U')a(T)$ *for all $U \in \text{SL}(n, \mathbb{Z}), U \equiv E \mod q$.*

The spaces $\mathbf{P}$ are filtred with respect to r, i.e.

$$\mathbf{P}(n, r-1, q, \varrho_0) \subset \mathbf{P}(n, r, q, \varrho_0).$$

We are more interested in the successive quotients than in the spaces $\mathbf{P}$ itself.

1.2 Notation.
$$\overline{\mathbf{P}} = \overline{\mathbf{P}(n,r,q,\varrho_0)} = \frac{\mathbf{P}(n,r,q,\varrho_0)}{\mathbf{P}(n,r-1,q,\varrho_0)}.$$

The so called "big singular space" is a certain subspace of $\overline{\mathbf{P}}$. For its definition we have to introduce a certain group.

1.3 Notation. *Let*
$$\Gamma = \Gamma_{n,\vartheta}(r)$$

be the subgroup of the Siegel modular group
$$\Gamma_n = \mathrm{Sp}(n,\mathbb{Z}),$$

which is generated by

a)
$$\begin{pmatrix} U' & 0 \\ 0 & U^{-1} \end{pmatrix}, \quad U \in \mathrm{SL}(n,\mathbb{Z}),$$

b)   *the "imbedded involution"*
$$I_r = \begin{pmatrix} E_r & E - E_r \\ E_r - E & E_r \end{pmatrix}, \quad E_r = E_r^{(n)} = \begin{pmatrix} E^{(r)} & 0 \\ 0 & 0 \end{pmatrix},$$

i.e.
$$I_r = \begin{pmatrix} E^{(r)} & 0 & 0 & 0 \\ 0 & 0 & 0 & E^{(n-r)} \\ 0 & 0 & E^{(r)} & 0 \\ 0 & -E^{(n-r)} & 0 & 0 \end{pmatrix}.$$

1.4 Definition. *The space*
$$\mathbf{M} = \mathbf{M}(n,r,q,\varrho_0)$$

consists of all holomorphic functions
$$f : \mathbb{H}_n \longrightarrow \mathcal{Z},$$

such that the function
$$Z \longmapsto \det{(CZ+D)}^{-r/2}\varrho_0(CZ+D)^{-1}f(M\langle Z\rangle)$$

is contained in the space $\mathbf{P}(1.1)$ for all
$$M = \begin{pmatrix} A & B \\ C & D \end{pmatrix} \in \Gamma_{n,\vartheta}(r) \qquad (1.3).$$

1.5 Notation. *We denote by*

$$\overline{\mathbf{M}} = \overline{\mathbf{M}(n, r, q, \varrho_0)}$$

the image of $\mathbf{M}$ *in the space* $\mathbf{P}$.

The condition formulated in 1.4 does not depend on the choice of a square root of $\det(CZ + D)$.

At a first glance the space $\overline{\mathbf{M}}$ looks rather big. But we shall see that it is finite dimensional.

Examples of the space M

We consider theta series of the type

$$\sum_{G = G^{(r,n)} \text{ integral}} P(S^{1/2} G) \exp \frac{\pi i}{q} \sigma \{S[G]Z + 2V'G\}.$$

Here $S = S^{(r)}$ is a positive definite symmetric matrix, V is a matrix of type $V = V^{(r,n)}$ and P is a polynomial,

$$P : \mathbb{C}^{(r,n)} \longrightarrow \mathcal{Z}.$$

Under certain conditions this theta series is contained in the space M. We are going to formulate those conditions.

1.6 Definition.. *Let* $S = S^{(r)}$ *be a positive definite symmetric integral matrix such that* $q^2 S^{-1}$ *is half integral if* $n = r + 1$ *and integral if* $n > r + 1$.

1) *An integral matrix* $H = H^{(r,n-r)}$ *is called* isotropic *with respect to* S, *if*

$$S^{-1}[H + qX]$$

is integral for all integral $X = X^{(r,n-r)}$.

2) *An integral matrix* $V = V^{(r,n)}$ *is called* isotropic, *if for every matrix* $U \in \mathrm{GL}(n, \mathbb{Z})$, *the matrix* H *defined by*

$$VU = (*, H), \qquad H = H^{(r,n-r)},$$

is isotropic in the sense of 1).

Remark.
Because of the conditions on S, a matrix $H = H^{(r,n-r)}$ is isotropic if and only if

$$S^{-1}[H] \quad \text{and} \quad \begin{cases} qS^{-1}H & \text{if } n > r + 1 \\ 2qS^{-1}H & \text{if } n = r + 1 \end{cases}$$

are both integral.

1.7 Proposition. *Assume that*

a) $S = S^{(r)}$ *is an integral positive symmetric matrix such that* $q^2 S^{-1}$ *is half integral if* $n = r + 1$ *and integral if* $n > r + 1$.

b) $V = V^{(r,n)}$ *is isotropic (1.6).*

c)   *P is a harmonic form with respect to* ϱ_0.

Then the above theta series is contained in the space **M**.

The proof is given in section 2.

Problem. *Is* $\overline{\mathbf{M}}$ *generated by the (images of) theta series described in 1.7?*

In this paper we formulate an elementary statement, which –if it is true– gives an affirmative answer to the problem. The lemma has the advantage that for given n, r, q it can be decided in a finite number of steps, i.e. in principle by a computer. Unfortunately it could not be proved or disproved in full generality up to now.

A complete proof will be given in the case $n \geq 2r$ (instead of $n > r$).

Why are we interested in this problem? To answer this, let

$$\Gamma_n[q, 2q] \subset \mathrm{Sp}(n, \mathbb{Z})$$

be Igusa's congruence subgroup (s.I sec.1). This group is a normal subgroup of the theta group,

$$\Gamma_{n,\vartheta} = \Gamma_n[1, 2] \ .$$

We consider modular forms with respect to a multiplier system v and with respect to

$$\varrho(A) := \det(A)^{r/2} \varrho_0(A).$$

An arbitrary element $N \in \Gamma_{n,\vartheta}$ of the theta group defines an isomorphism

$$[\Gamma_n[q, 2q], \varrho, v] \to [\Gamma_n[q, 2q], \varrho, v^N]$$

$$f(Z) \mapsto \det(CZ + D)^{-r/2} \varrho_0(CZ + D)^{-1} f(N\langle Z\rangle), \quad N = \begin{pmatrix} A & B \\ C & D \end{pmatrix},$$

where v^N denotes the "conjugate" multiplier system of v. (We use the notations of I sec.3).

During the rest of this book we make the

1.8 Assumption.

$$v^N(M) \cdot \det(CZ + D)^{-r/2} = 1 \quad \text{for all } N \in \Gamma_{n,\vartheta},$$

if

$$M = \begin{pmatrix} A & B \\ C & D \end{pmatrix} \in \Gamma_n[q, 2q], \quad C = 0, \det D = 1 \ .$$

For example the trivial multiplier system $v \equiv 1$ if r is even and the r−th power of the theta multiplier system v_ϑ have this property.

1.9 Remark.. *If the assumption 1.8 for the multiplier system is satisfied, we have*

a) $$[\Gamma_n[q, 2q], \varrho, v] \subset \mathbf{M},$$

b) $$[\Gamma_n[q, 2q], \varrho, v] \cap \mathbf{P}(n, r - 1, q, \varrho_0) = 0 .$$

The proof depends on two facts.

1)   Each non-vanishing modular form $f \in [\Gamma_n[q, 2q], \varrho, v]$ is **singular**, i.e. it satisfies condition b) in 1.1, but not for $r - 1$ instead of r (III 4.4).

2)   The congruence group $\Gamma_n[q, 2q]$ is a normal subgroup of the theta group

$$\Gamma_{n,\vartheta} = \Gamma_n[1, 2].$$

This group acts on our spaces of modular forms. (Here our assumption about the multiplier systems gets important.) On the other hand the group $\Gamma_{n,\vartheta}(r)$ defined in 1.3 is contained in the theta group. This gives us $f \in \mathbf{M}$.

Hence an affirmative answer to our problem would imply that each modular form in $[\Gamma_n[q, 2q], \varrho, v]$ can be written as linear combination of certain **finitely many** theta series.

In case $n > r+1$ those theta series are actually contained in $[\Gamma_n[q, 2q], \varrho, v_S]$ for a certain v_S depending on S (s.II sec.7). So in the case $n > r + 1$:

$$[\Gamma_n[q, 2q], \varrho, v] = \text{span of theta series 1.7 with } v = v_S.$$

We should mention that ENDRES has proved this in the case $r = 1$ and for the trivial one dimensional representation ($\mathcal{Z} = \mathbb{C}$) [En].

We should also mention that the theory gets far easier if one considers only congruence groups, which contain **all** unimodular substitutions

$$Z \longmapsto Z[U], \quad U \in \mathrm{GL}(n, \mathbb{Z}) .$$

In this case the characteristics V occuring can taken to be 0. For details we refer to the papers [AZ], [En], [Fr3].

There is another general result of Howe [Ho], who proved by means of representation theory that each singular modular form with even r can be represented as a linear combination of theta series. But he needs infinitely many theta series. For example he has to admit all theta series with **rational** characteristic V and it is not at all clear how to exhibit a finite generating system. But this is necessary, if one wants to determine the dimensions of the spaces $[\Gamma_n[q, 2q], \varrho, v]$. In some cases such dimensions have geometric meaning (s. [Wel], [AZ]). This is one reason, why we try to refine Howe's results.

One difficulty of our problem arises from the following fact:
There exist linear relations between the theta series described in 1.7 for fixed S and varying characteristic V. For example the classical **Riemann theta relations** are of this type. It will turn out that our problem is connected with the theory of theta relations of "Riemannian type".

One may ask whether the big space $\overline{M}$ is of own interest or whether only spaces of modular forms should be considered. The advantage of $\overline{M}$ is that its definition does not depend on questions of multiplier systems or of choices of suitable congruence subgroups. The high significance of the "imbedded involution" for the whole theory is processed into the definition of $\overline{M}$. It is very likely that the concept of the big space carries over to more general modular groups which can be attached to certain involutive algebras and that our theory can be generalized to those cases.

2   Theta series contained in M (The proof of Proposition 1.7)

In the case $n \geq r+2$ proposition 1.7 is a consequence of the fact that the theta series, which we consider, are modular forms on $\Gamma_n[q, 2q]$. This was one of the main results of the general theta transformation formalism. We nevertheless include a complete proof which covers also the case $n = r + 1$. The reason is that –after the space $\overline{M}$ has been defined– things get far easier, because one is reduced to deal with only one simple modular substitution, namely the imbedded involution. The general transformation formalism in some sense is only necessary to justify the definition of $\overline{M}$.

In this and the next section we use the modified notation

$$\vartheta_{S,P}(Z; U, V) := \vartheta_P \begin{bmatrix} U \\ V \end{bmatrix} (S; Z),$$

which is advantegeous if the expression for the characteristik is long. We have to determine the action of the imbedded involution I_r (s.1.3) on a theta series

$$f(Z) = \sum_{G=G^{(r,n)}} P(S^{1/2}G) \exp \frac{\pi i}{q} \sigma \{S[G]Z + 2V'G\}$$

For this purpose we decompose the matrix $Z \in \mathbb{H}_n$ into 4 blocs:

$$Z = \begin{pmatrix} Z_0 & Z_1 \\ Z_1' & Z_2 \end{pmatrix}; \quad Z_0 = Z_0^{(r)}.$$

The imbedded involution I_r acts as

$$I_r(Z) = \begin{pmatrix} Z_0 - Z_2^{-1}[Z_1'] & -Z_1 Z_2^{-1} \\ -Z_2^{-1} Z_1' & -Z_2^{-1} \end{pmatrix}.$$

If we decompose

$$G = (G_1, G_2); \quad G_1 = G_1^{(r,r)}, G_2 = G_2^{(r,n-r)}$$

and analogous $V = (V_1, V_2)$, we obtain

$$f(Z) = \sum_{G_1} \exp \frac{\pi i}{q} \sigma\{S[G_1]Z_0 + 2V_1'G_1\} h_{G_1}(Z_2, Z_1),$$

where

$$h_{G_1}(Z_2, Z_1)$$
$$= \sum_{G_2} P(S^{1/2}G_1, S^{1/2}G_2) \exp \frac{\pi i}{q} \sigma\{S[G_2]Z_2 + 2G_1'SG_2Z_1' + 2V_2'G_2\}.$$

We have to investigate the function

$$g(Z) = \det(Z/i)^{-r/2} \varrho_0 \begin{pmatrix} E & 0 \\ -Z_2^{-1}Z_1' & -Z_2^{-1} \end{pmatrix} f(I_r\langle Z\rangle).$$

We now fix G_1 for a while and write

$$h(Z_2, Z_1) = h_{G_1}(Z_2, Z_1),$$

$$P_0(G_2) = P(S^{1/2}G_1, q^{1/2}G_2).$$

The function h can be written as theta series with characteristics

$$h(Z_2, Z_1) = \vartheta_{S/q, P_0}(Z_2; 0, (V_2 + SG_1Z_1)/q).$$

To determine $g(Z)$ we have to substitute

$$(Z_2, Z_1) \longmapsto (-Z_2^{-1}, -Z_1Z_2^{-1}).$$

By means of the transformation formula II 5.5 one obtains

$$h(-Z_2^{-1}, -Z_1Z_2^{-1}) = \vartheta_{S/q, P_0}(-Z_2^{-1}; 0, (V_2 - SG_1Z_1Z_2^{-1})/q) =$$
$$\det(S/q)^{-(n-r)/2} \det(Z_2/i)^{r/2} \cdot \vartheta_{qS^{-1}, Q_0}(Z_2; SG_1Z_1Z_2^{-1} - V_2, 0),$$

where $Q_0(H, Z_2)$ is the Gauss transform of the polynomial

$$U = U^{(r,n-r)} \longmapsto P_0(U(Z_2/i)^{1/2})$$

at

$$-iH(Z_2/i)^{1/2}.$$

We have

$$P_0(U(Z/i)^{1/2}) = P(S^{1/2}G_1, q^{1/2}U(Z/i)^{1/2}).$$

We now assume that P is a harmonic form II 6.5. It follows that the above function is harmonic in the variable U. Hence it equals its Gauss transform.

$$Q_0(H, Z_2) = P(S^{1/2}G_1, -q^{1/2}HZ_2)$$

$$= \varrho \begin{pmatrix} E & 0 \\ 0 & -Z_2 \end{pmatrix} P(S^{1/2}G_1, q^{1/2}H).$$

Hence

$$\det(S/q)^{(n-r)/2}g(Z) = \varrho_0 \begin{pmatrix} E & 0 \\ -Z_2^{-1}Z_1' & -Z_2^{-1} \end{pmatrix} \varrho_0 \begin{pmatrix} E & 0 \\ 0 & -Z_2 \end{pmatrix}$$

$$\cdot \sum_{G_1} \exp \frac{\pi i}{q} \sigma\{S[G_1](Z_0 - Z_2^{-1}[Z_1']) + 2V_1'G_1\}$$

$$\cdot \vartheta_{qS^{-1}, P_0}(Z_2; SG_1Z_1Z_2^{-1} - V_2, 0).$$

A little calculation shows

$$\varrho_0 \begin{pmatrix} E & 0 \\ -Z_2^{-1}Z_1' & -Z_2^{-1} \end{pmatrix} \varrho_0 \begin{pmatrix} E & 0 \\ 0 & -Z_2 \end{pmatrix}$$

$$\cdot P_0\big((qS^{-1})^{1/2}(G_2 + (SG_1Z_1Z_2^{-1} - V_2)/q)\big)$$

$$= \varrho_0 \begin{pmatrix} E & 0 \\ -Z_2^{-1}Z_1' & E \end{pmatrix} \cdot P_0\big((qS^{-1})^{1/2}(G_2 + (SG_1Z_1Z_2^{-1} - V_2)/q)\big)$$

$$= P\big(S^{1/2}G_1, qS^{-1/2}(G_2 - V_2/q)\big),$$

hence

$$\det(S/q)^{(n-r)/2}g(Z) = \sum_{G=(G_1,G_2)} P\big(S^{1/2}G_1, qS^{-1/2}(G_2 - V_2/q)\big) \exp \frac{\pi i}{q} \sigma(R),$$

where R is given by

$$R = S[G_1](Z_0 - Z_2^{-1}[Z_1']) + 2V_1'G_1 +$$
$$q^2 S^{-1}[G_2 + (SG_1Z_1Z_2^{-1} - V_2)]Z_2.$$

A simple calculation yields

$$\sigma(R) = \sigma\{S[G_1, qS^{-1}(G_2 - V_2/q)]Z + 2V_1'G_1\}$$

and hence the final formula for g

$$\det(S/q)^{(n-r)/2}g(Z) =$$
$$\sum_G P(S^{1/2}G_1, qS^{-1/2}(G_2 - V_2/q)) \cdot \exp \frac{\pi i}{q} \sigma\{S[G_1, qS^{-1}(G_2 - V_2/q)]Z + 2V_1'G_1\}.$$

We now consider the space

$$\Theta = \Theta(n, r, q, \varrho_0),$$

which is generated by all those theta series, where (P, S, V) satisfies the conditions of proposition 1.7. We obviously have

a) $\Theta \subset \mathbf{P}$ (s.1.1).

b)   The group $SL(n, \mathbb{Z})$ acts on Θ.

2.1 Lemma. *If f is in* Θ, *then g is also in* Θ.

Of course proposition 1.7 is aconsequence of a), b) and the lemma.

Proof of 2.1. The proof depends on the following

2.2 Remark. *If* $V = V^{(r,n)}$ *is an isotropic matrix, then*

$$VU; \quad U = U^{(n,n)} \in \mathbb{Z}^{(n,n)}$$

is isotropic for all integral U (not only for unimodular ones as demanded in 1.6).

The proof is an immediate consequence of the remark which follows definition 1.6.

Corollary. *Let* $\mathcal{L}(V)$ *be the subgroup of* $\mathbb{Z}^r = \mathbb{Z}^{(r,1)}$ *generated by* $q\mathbb{Z}^r + S\mathbb{Z}^r$ *and by the columns of V. Any matrix in* $\mathbb{Z}^{(r,n)}$ *whose columns are contained in* $\mathcal{L}(V)$ *is isotropic.*

2.3 Definition. *A group* $\mathcal{L}$

$$q\mathbb{Z}^r + S\mathbb{Z}^r \subset \mathcal{L} \subset \mathbb{Z}^r$$

is called isotropic with respect to S, if any matrix $V \in \mathbb{Z}^{(r,n)}$ *whose columns are contained in* $\mathcal{L}$, *is isotropic (in the sense of 1.6).*

The "orthogonal complement" of the isotropic group $\mathcal{L}$ is defined by

$$\mathcal{K} = \{h \in \mathbb{Z}^r; \ h'x \equiv 0 \ \mathrm{mod}\, q \quad \text{for all } x \in \mathcal{L}\}.$$

The functions

$$\varphi_V(G) = \exp \frac{2\pi i}{q} \sigma(G'V), \quad \mathcal{L}(V) \subset \mathcal{L},$$

depend only on G mod $\mathcal{K}^n$. Hence we may consider them as functions

$$\varphi_V : \mathbb{Z}^{(r,n)}/\mathcal{K}^n \longrightarrow \mathbb{C}.$$

From the "Fourier analysis" of the finite Abelian group $\mathbb{Z}^{(r,n)}/\mathcal{K}^n$ it follows that each function on this group is a linear combination of characters. This gives us

2.4 Remark. *Let* $\mathcal{L} \subset \mathbb{Z}^r$ *be an isotropic group (2.3) and*

$$\varphi : \mathbb{Z}^{(r,n)}/\mathcal{K}^n \longrightarrow \mathbb{C}$$

an arbitrary function. The series

$$\sum_{G \in \mathbb{Z}^{(r,n)}} \varphi(G) P(S^{1/2}G) \exp \frac{\pi i}{q} \sigma(S[G])$$

is contained in the space Θ.

The isotropic space $\mathcal{L}(V)$ with respect to the given V may be written as

$$\mathcal{L} = \mathcal{L}(V) = A\mathbb{Z}^r.$$

The matrices

$$A; \; qA^{-1} \text{ and } A^{-1}S$$

are integral. Of course the isotropy of V implies that

$$\tilde{S} := S^{-1}[A]$$

is also integral. The series g can obviously be rewritten as

$$\sum \varphi(G)\tilde{P}(\tilde{S}^{1/2}G) \exp \frac{\pi i}{q}\sigma(\tilde{S}[G]),$$

$$\tilde{P}(X) = P(X),$$

where φ is the function

$$\varphi : \mathbb{Q}^{(r,n)} \longrightarrow \mathbb{C}$$

defined by

$$\varphi(X_1, X_2) = \exp \frac{\pi i}{q}\sigma(V_1'G_1)$$

if

$$G_1 := AS^{-1}X_1; \quad G_2 := qA^{-1}X_1$$

both are integral and

$$\varphi(X_1, X_2) = 0 \text{ elsewhere.}$$

We have to verify that g is of the type (2.4). This means

1) $\tilde{P}$ is harmonic.
This follows from the fact that the matrix $S^{1/2}A^{-1}\tilde{S}^{-1/2}$ is orthogonal.

2) $\tilde{S}$ satisfies the same conditions as S.
Actually from the integrality of the matrices A and S follows that $\tilde{S}$ and $q\tilde{S}^{-1}$ are integral.

3)   If G_1 and G_2 both are integral, than X_1 and X_2 are integral. We hence may consider φ as a function

$$\varphi : \mathbb{Z}^{(r,n)} \longrightarrow \mathbb{C}.$$

To complete the proof we have to construct an isotropic group $\tilde{\mathcal{L}}$ with respect to $\tilde{S}$ such that φ actually is defined on $\mathbb{Z}^{(r,n)}/\tilde{\mathcal{K}}^{(r,n)}$. ($\tilde{\mathcal{K}}$ denotes the orthogonal complement of $\tilde{\mathcal{L}}$). This means precisely

a) $AS^{-1}\tilde{\mathcal{K}} \subset \mathcal{K}$,
b) $qA^{-1}\tilde{\mathcal{K}} \subset \mathbb{Z}^r$.

We can take

$$\tilde{\mathcal{L}} := A'\mathbb{Z}^r + \tilde{S}\mathbb{Z}^r.$$

This group is isotropic with respect to $\widetilde{S}$ because

$$\widetilde{S}^{-1}[A'] = S[A'^{-1}][A'] = S$$

is integral (and because $q\widetilde{S}^{-1}$ is integral). From

$$\widetilde{\mathcal{L}} \supset A'\mathbb{Z}^r$$

follows

$$\widetilde{\mathcal{K}} \subset qA^{-1}\mathbb{Z}^r,$$

i.e. b). From

$$\widetilde{\mathcal{L}} \supset \widetilde{S}\mathbb{Z}^r$$

follows

$$\widetilde{\mathcal{K}} \subset qA^{-1}SA'^{-1}\mathbb{Z}^r$$

or

$$AS^{-1}\widetilde{\mathcal{K}} \subset qA'^{-1}\mathbb{Z}^r = \mathcal{K}.$$

3   Fourier-Jacobi expansion

As in the last section, we decompose the variable $Z \in \mathbb{H}_n$ into 4 blocs

$$Z = \begin{pmatrix} Z_0 & Z_1 \\ Z_1' & Z_2 \end{pmatrix}; \quad Z_0 = Z_0^{(r)}.$$

The Fourier-Jacobi expansion of an element

$$f(Z) = \sum a(T) \exp \frac{\pi i}{q} \sigma(TZ)$$

of our space $\mathbf{P}$ (1.1) is defined as

$$f(Z) = \sum_{T_0 = T_0' \geq 0} \varphi_{T_0}(Z_1, Z_2) \exp \frac{\pi i}{q} \sigma(T_0 Z_0) \quad (T_0 = T_0^{(r)}),$$

where

$$\varphi_{T_0}(Z_1, Z_2) = \sum_T a(T) \exp \frac{\pi i}{q} \sigma\{T_2 Z_2 + 2T_1' Z_1\}, \quad T = \begin{pmatrix} T_0 & T_1 \\ T_1' & T_2 \end{pmatrix},$$

is the so called Fourier-Jacobi coefficient with respect to the matrix T_0. This coefficient is a holomorphic function on $\mathbb{C}^{(r,n-r)} \times \mathbb{H}_{n-r}$. We are going to write this coefficient as a linear combination of theta series. For this purpose we need

3.1 Remark. *Let $S = S^{(r)} > 0$ be a rational positive symmetric matrix. The set of all rational matrices*

$$T = T^{(n)} = T' = \begin{pmatrix} S & * \\ * & * \end{pmatrix}$$

with the property

$$T \geq 0 \quad and \quad \operatorname{rank} T \leq r$$

is in one to one correspondence with the set of all rational matrices

$$G = G^{(r,n-r)}$$

by means of the correspondence

$$T = S[E^{(r)}, S^{-1}G] = \begin{pmatrix} S & G \\ G' & S^{-1}[G] \end{pmatrix}.$$

Proof. The conditions

$$T \geq 0, \quad \operatorname{rank} T \leq r$$

imply the existence of a representation

$$T = \begin{pmatrix} T_0^{(r)} & 0 \\ 0 & 0 \end{pmatrix} [U], \quad U \in \mathrm{GL}(n, \mathbb{Q})$$

or

$$T = T_0[A, B], \quad A = A^{(r)}, \quad B = B^{(r,n-r)}.$$

This implies 3.1.

Because of 3.1 the Fourier-Jacobi coefficient of an integral matrix $T_0 = S = S^{(r)} > 0$ is of the form

$$\varphi_S(Z_1, Z_2) = \sum a(S[E, S^{-1}G]) \cdot \exp \frac{\pi i}{q} \sigma \{ S^{-1}[G]Z_2 + 2G'Z_1 \}.$$

The summation is taken over all

$$G = G^{(r,n-r)} \text{ integral and } S^{-1}[G] \text{ integral.}$$

If

$$G \in \mathbb{Z}^{(r,n-r)}/q S \mathbb{Z}^{(r,n-r)}$$

runs through a system of representatives mod qS, we obtain

$$\varphi_S(Z_1, Z_2) = \sum_{G \bmod qS} \sum_{H \text{ integral}} \varrho_0 \begin{pmatrix} E & 0 \\ qH' & E \end{pmatrix} a(S[E, S^{-1}G]) \cdot$$

$$\exp \frac{\pi i}{q} \{ S^{-1}[G + qSH]Z_2 + 2(G + qSH)'Z_1 \}.$$

The inner sum can be written as theta series with characteristic and polynomial coefficients. With the notations from sec.2 we obtain:

$$\varphi_S(Z_1, Z_2) = \sum_{G \bmod qS} \vartheta_{qS,Q_G}(Z_2; q^{-1}S^{-1}G, SZ_1).$$

Here the polynomial Q_G is defined as

$$Q_G((qS)^{1/2}(X + q^{-1}S^{-1}G)) = \varrho_0 \begin{pmatrix} E & 0 \\ qX' & E \end{pmatrix} a(S[E, S^{-1}G]),$$

i.e.

$$Q_G(X) = \varrho_0 \begin{pmatrix} E & 0 \\ q^{1/2}X'S^{-1/2} & E \end{pmatrix} \cdot a_G,$$

where

$$a_G := \varrho_0 \begin{pmatrix} E & 0 \\ -G'S^{-1} & E \end{pmatrix} a(S[E, S^{-1}G]).$$

Now we consider the imbedded involution

$$I_r = \begin{pmatrix} E_r & E - E_r \\ E_r - E & E_r \end{pmatrix}, \quad E_r = \begin{pmatrix} E^{(r)} & 0 \\ 0 & 0 \end{pmatrix},$$

which acts as

$$I_r\langle Z \rangle = \begin{pmatrix} Z_0 - Z_2^{-1}[Z_1'] & -Z_1 Z_2^{-1} \\ -Z_2^{-1}Z_1' & -Z_2^{-1} \end{pmatrix}.$$

Next we compute the function

$$g(Z) := \det(Z_2/i)^{-r/2} \varrho_0 \begin{pmatrix} E & 0 \\ -Z_2^{-1}Z_1' & -Z_2^{-1} \end{pmatrix} f(I_r\langle Z \rangle).$$

We apply the transformation formula to the theta functions which occur in the Fourier-Jacobi expansion of f:

$$\varphi_S(-Z_1 Z_2^{-1}, -Z_2^{-1})$$
$$= \sum_{G \bmod qS} \exp 2\pi i \sigma((q^{-1}S^{-1}G)'S(-Z_1 Z_2^{-1}))$$
$$\det(qS)^{(r-n)/2} \det(Z_2/i)^{r/2} \vartheta_{(qS)^{-1}, \widetilde{Q}_G}(Z_2; SZ_1 Z_2^{-1}, q^{-1}S^{-1}G)$$

$$= \det(qS)^{(r-n)/2} \det(Z_2/i)^{r/2} \cdot \sum_{G \bmod qS}$$
$$\sum_{H \text{ integral}} \widetilde{Q}_G((qS)^{-1/2}(H + SZ_1 Z_2^{-1}), Z_2)$$
$$\exp \frac{\pi i}{q} \sigma\{S^{-1}[H + SZ_1 Z_2^{-1}]Z_2 + 2H'S^{-1}G\}.$$

By definition $\tilde{Q}_G(R, Z_2)$ is the Gauß transform of
$$U \longmapsto Q_G(U(Z_2/i)^{1/2})$$
at
$$-iR \cdot (Z_2/i)^{1/2}.$$
This equals
$$\int Q_G(-RZ_2 + X(Z_2/i)^{1/2}) \exp\{-\pi\sigma(X'X)\}dX$$

$$= \varrho_0 \begin{pmatrix} E & 0 \\ -q^{1/2}Z_2 R'S^{-1/2} & E \end{pmatrix} \int Q_G(X(Z_2/i)^{1/2}) \exp\{-\pi\sigma(X'X)\}dX.$$

Now we obtain for g an expansion
$$g(Z) = \sum \psi_{T_0}(Z_1, Z_2) \exp\frac{\pi i}{q}\sigma(T_0 Z_0),$$
where
$$\psi_{T_0}(Z_1, Z_2) =$$
$$\det(Z_2/i)^{-r/2}\varrho_0 \begin{pmatrix} E & 0 \\ -Z_2^{-1}Z_1' & -Z_2^{-1} \end{pmatrix} \exp\frac{-\pi i}{q}\sigma(T_0 Z_2^{-1}[Z_1'])\varphi_{T_0}(-Z_1 Z_2^{-1}, Z_2),$$
hence
$$\psi_S(Z_1, Z_2) = (\det qS)^{(r-n)/2}\varrho_0 \begin{pmatrix} E & 0 \\ -Z_2^{-1}Z_1' & -Z_2^{-1} \end{pmatrix}$$

$$\sum_{G \bmod qS} \sum_{H \text{ integral}} \tilde{Q}_G((qS)^{-1/2}(H + SZ_1 Z_2^{-1}), Z_2)$$

$$\exp\frac{\pi i}{q}\sigma\{S^{-1}[H]Z_2 + 2H'(Z_1 + S^{-1}G)\}.$$

We now evaluate $\tilde{Q}_G(R, Z_2)$ at $R = (qS)^{-1/2}(H + SZ_1 Z_2^{-1})$ and multiply from the left by
$$\varrho_0 \begin{pmatrix} E & 0 \\ -Z_2^{-1}Z_1' & -Z_2^{-1} \end{pmatrix}.$$
Because of
$$\begin{pmatrix} E & 0 \\ -Z_2^{-1}Z_1' & -Z_2^{-1} \end{pmatrix}\begin{pmatrix} E & 0 \\ -q^{1/2}Z_2 R'S^{-1/2} & E \end{pmatrix} = \begin{pmatrix} E & 0 \\ H'S^{-1} & E \end{pmatrix}\begin{pmatrix} E & 0 \\ 0 & -Z_2^{-1} \end{pmatrix}$$
we obtain
$$\psi_S(Z_1, Z_2) =$$
$$(\det S)^{(r-n)/2} \sum_{G \bmod qS} \sum_{H \text{ integral}} \varrho_0 \begin{pmatrix} E & 0 \\ H'S^{-1} & E \end{pmatrix}\varrho_0 \begin{pmatrix} E & 0 \\ 0 & -Z_2^{-1} \end{pmatrix}$$
$$\int Q_G(X(Z_2/i)^{1/2}) \exp\{-\pi\sigma(X'X)\}dX$$
$$\exp\frac{\pi i}{q}\sigma\{S^{-1}[H]Z_2 + 2H'(Z_1 + S^{-1}G)\}.$$

This is a Fourier series in the variable Z_1. For the rest of this section we make the

Assumption. *The functions f and g are contained in the space* **P**.

This implies that g is periodic as a function of Z_2. Hence we know that the Fourier coefficients of

$$Z_1 \longmapsto \psi_S(Z_1, Z_2)$$

are periodic functions of Z_2 .

We now put

$$A_H = \sum_{G \bmod qS} \varrho_0 \begin{pmatrix} E & 0 \\ -G'S^{-1} & E \end{pmatrix} a(S[E, S^{-1}G]) \exp \frac{2\pi i}{q} \sigma(H'S^{-1}G)$$

and

$$Q^a(X) := \varrho_0 \begin{pmatrix} E & 0 \\ X'S^{-1/2} & E \end{pmatrix} a \quad \text{for } a \in \mathcal{Z}$$

We obtain:

Let a be a linear combination of the vectors A_H. The function

$$Z_2 \longmapsto \varrho_0 \begin{pmatrix} E & 0 \\ 0 & -Z_2^{-1} \end{pmatrix} \int Q^a(q^{-1/2} X(Z_2/i)^{1/2}) \exp\{-\pi\sigma(X'X)\} dX$$

is periodic. On the other hand it is a rational function of $(Z_2/i)^{1/2}$. Therefore it has to be constant! We will use this in the special case $Z_2 = t \cdot iE, t > 0$.

This property is rather strong. For example we will derive from it that $Q(X) := Q^a(X)$ is harmonic!

For this purpose we consider also translates of Q:

$$Q_t(X) := Q(tX) = \varrho_0 \begin{pmatrix} E & 0 \\ tX'S^{-1/2} & E \end{pmatrix} a \quad (t \in \mathbb{C}).$$

Its Gauss transform is

$$Q_t^*(X) = \varrho_0 \begin{pmatrix} E & 0 \\ tX'S^{-1/2} & E \end{pmatrix} \int \varrho_0 \begin{pmatrix} E & 0 \\ tU'S^{-1/2} & E \end{pmatrix} a \cdot \exp\{-\pi\sigma(U'U)\} dU$$

$$= \varrho_0 \begin{pmatrix} E & 0 \\ tX'S^{-1/2} & E \end{pmatrix} Q_t^*(0).$$

Now a simple computation using the above invariance property gives us

$$Q_t^*(tX) = \varrho_0 \begin{pmatrix} E & 0 \\ 0 & t^2 E \end{pmatrix} Q^*(X).$$

On the other hand

$$\lim_{t \to 0} Q_t^*(t^{-1}X) = \lim_{t \to 0} \int Q(tU + X) \exp -\pi\sigma(U'U) dU = Q(X).$$

We obtain

$$\lim_{t \to 0} \varrho_0 \begin{pmatrix} E & 0 \\ 0 & t^2 E \end{pmatrix} Q^*(t^{-2}X) = Q(X).$$

The vector space $\mathcal{Z}$ admits a basis such that $\varrho_0 \begin{pmatrix} E & 0 \\ 0 & t^2 E \end{pmatrix}$ acts on each basis element by multiplication with a certain power of t. An easy consequence of II 1.3 is, that each component of Q with respect to this basis is a homogeneous polynomial. Now from II 1.3 follows that the corresponding component of Q^* is homogeneous too. From II 1.3, we obtain that Q is harmonic and that $Q = Q^*$. What we have proved is the important

3.2 Lemma. *Assume that g as well as f are contained in the space* $\mathbf{P}$. *Let a be an element of the subspace* $\mathcal{Z}_0$ *of* $\mathcal{Z}$, *which is generated by the vectors*

$$a_G := \varrho_0 \begin{pmatrix} E & 0 \\ -G'S^{-1} & E \end{pmatrix} a(S[E, S^{-1}G]) \quad (G \text{ and } S^{-1}[G] \text{ integral})$$

(or equivalently A_H). Then the polynomial

$$Q^a(X) = \varrho_0 \begin{pmatrix} E & 0 \\ X'S^{-1/2} & E \end{pmatrix} a$$

is a harmonic form with respect to the representation

$$A = A^{(n-1)} \longmapsto \varrho_0 \begin{pmatrix} E & 0 \\ 0 & A \end{pmatrix}.$$

One furthermore has

$$\varrho_0 \begin{pmatrix} E & A_2 \\ 0 & A_4 \end{pmatrix} a = a \quad \text{for all } a \in \mathcal{Z}_0.$$

Thanks to 3.2 we obtain a cosiderable simplification of the formula for the Fourier-Jacobi coefficient:

$$\psi_S(Z_1, Z_2) =$$
$$(\det S)^{(r-n)/2} \sum_{H \text{ integral}} \varrho_0 \begin{pmatrix} E & 0 \\ H'S^{-1} & E \end{pmatrix} A_H \exp \frac{\pi i}{q} \sigma\{S^{-1}[H]Z_2 + 2H'Z\}.$$

If we write the Fourier series of g in the form

$$g(Z) = \sum b(T) \exp \frac{\pi i}{q} \sigma(TZ),$$

we obtain

3.3 Proposition. *Put*

$$\alpha(E, G) := \varrho_0 \begin{pmatrix} E & 0 \\ -G' & E \end{pmatrix} a(S[E, G])$$

$$\beta(E, H) := \varrho_0 \begin{pmatrix} E & 0 \\ -H' & E \end{pmatrix} b(S[E, H])$$

$$:= 0, \quad \text{if } S[E, H] \text{ is not integral}$$

Then the inversion formula

$$\beta(E, H) = (\det S)^{(r-n)/2} \sum_{G \bmod q} \alpha(E, G) \exp \frac{2\pi i}{q} \sigma(G'SH)$$

holds. Here $H = H^{(r,n-r)}$ is a matrix such that SH is integral. The summation is taken over a maximal system of mod q different matrices G, such that $S[E, G]$ is integral.

This inversion formula was discovered in special cases by Endres [En].

Remark: The mapping $f \mapsto g$ being involutive, we may interchange the roles of α and β in the inversion formula!

The harmonic polynomial $Q^a(X), X = X^{(r,n-r)}$ can not yet be used as coefficient for theta series with respect to the representation ϱ_0, because it depends on too few variables. Hence we have to blow it up to a polynomial $P^a(X), X = X^{(r,n)}$. We define:

$$P^{(}S^{1/2}X) = \varrho_0 \begin{pmatrix} X_1' & 0 \\ X_2' & E \end{pmatrix} E(S)^{-1}a \quad (a \in \mathcal{Z}_0)$$
$$X = (X_1^{(r)}, X_2^{(r,n-r)}).$$

Here $E(S)$ denotes the order of the unit group

$$\mathcal{E}(S) = \{U \in \mathrm{GL}(r, \mathbb{Z}) \mid S[U] = S\}.$$

3.4 Proposition. *For each vector $a \in \mathcal{Z}_0$ (s.3.2) the function $P^a(X)$ is a harmonic form with respect to the representation ϱ_0. It is characterized by*

$$P^a(S^{1/2}, 0) = E(S)^{-1}a.$$

Proof. From 3.2, it follows

$$P^a(XA) = \varrho_0(A')P^a(X) \quad \text{for all } A \in \mathrm{GL}(n, \mathbb{C}).$$

It remains to prove that P^a is harmonic. By construction of P we have

$$P^a(X_1, X_2) = E(S)^{-1}\varrho_0 \begin{pmatrix} X_1' & 0 \\ 0 & E \end{pmatrix} Q^a(X_2).$$

Therefore P^a is harmonic as function of an arbitrary column of X_2. As we can permute the columns of X by multiplication

$$X \longmapsto XA, \quad A \text{ suitable,}$$

$P^a(X)$ is harmonic as function of each column of X. This of course implies that $P^a(X)$ is harmonic.

4   The hidden relations

We consider a function

$$f(Z) = \sum a(T) \exp \frac{\pi i}{q} \sigma(TZ)$$

from our space **P** (1.1). We fix a positive integral $S = S^{(r)}$ and investigate the system of Fourier coefficients

$$a(S[P]), \quad P \text{ primitive}.$$

(Primitive means that there exists a matrix $U = \begin{pmatrix} P \\ * \end{pmatrix} \in \mathrm{SL}(n, \mathbb{Z}/q\mathbb{Z}).$)

4.1 Remark. *Let f be a function in* **P** *and $S = S^{(r)} > 0$ a positive integral matrix. For each primitive $P = P^{(r,n)}$ we choose a matrix*

$$U = \begin{pmatrix} P \\ * \end{pmatrix} \in \mathrm{SL}(n, \mathbb{Z}).$$

The expression

$$\alpha(P) := \varrho_0(U'^{-1}) a\left(\begin{pmatrix} S & 0 \\ 0 & 0 \end{pmatrix} [U] \right)$$

does not depend on the choice of U. It depends only on P mod q.

Proof. The independence of the choice of U is a consequence of the formula

$$\begin{pmatrix} S & 0 \\ 0 & 0 \end{pmatrix} \begin{bmatrix} E & 0 \\ * & * \end{bmatrix} = \begin{pmatrix} S & 0 \\ 0 & 0 \end{pmatrix}$$

together with the invariance property 1.1, c). Both together show

$$\varrho_0(U'^{-1}) a \begin{pmatrix} S & 0 \\ 0 & 0 \end{pmatrix} = a \begin{pmatrix} S & 0 \\ 0 & 0 \end{pmatrix} \quad \text{for} \quad U = \begin{pmatrix} E & 0 \\ * & * \end{pmatrix}.$$

(This is a polynomial identity. Therefore the condition $U \equiv E$ mod q in 1.1 can be omitted.) It remains to show that $\alpha(P)$ only depends on P mod q. One has to show that for two **primitive** matrices P_1, P_2 , which are equivalent mod q a unimodular matrix U with the property

$$P_1 U = P_2, \quad U \equiv E \text{ mod } q$$

can be found. This means that for each primitive matrix P with the property $P \equiv (E, 0)$ mod q there exists a unimodular matrix U, $U \equiv E$ mod q, $U = \begin{pmatrix} P \\ * \end{pmatrix}$. This is easy to show and can be left to the reader.

Notation. *The space $\mathcal{P}$ consists of all primitive matrices $P = P^{(r,n)}$ of the ring $\mathbb{Z}/q\mathbb{Z}$.*

The natural map

$$\mathrm{SL}(n, \mathbb{Z}) \longrightarrow \mathrm{SL}(n, \mathbb{Z}/q\mathbb{Z})$$

being surjective, we may consider α (4.1) as a mapping

$$\alpha : \mathcal{P} \longrightarrow \mathcal{Z}.$$

We now formulate as an immediate consequence of the inversion formula 3.3 what we call

the hidden relations.

4.2 Proposition. *Assume that f is an element of the space* **M** *(1.4) and that $S = S^{(r)}$ is a positive integral matrix. Let furthermore W be an integral matrix such that $S^{-1}[W]$ is not integral. Then for all $U \in \mathrm{SL}(n, \mathbb{Z})$ the relation*

$$\sum \alpha((E, Y)U) \exp \frac{2\pi i}{q} \sigma(Y'W) = 0$$

holds. The summation has to be taken over a maximal system of $\mod q$ different matrices Y such that $S[E, Y]$ is integral.

Proof. The unimodular group acts on **M**. Hence it is sufficient to treat the case $U = E$. From the assumption on W, it follows that $\beta(E, S^{-1}H)$ in the inversion formula 3.3 vanishes.

We will specialize the hidden relations 4.2 to so-called f-kernel forms S:

4.3 Definition. *A positive integral matrix $S = S^{(r)}$ is called kernel form of a function*

$$f(Z) = \sum a(T) \exp \frac{\pi i}{q} \sigma(TZ) \quad \in \mathbf{P} \ (1.1),$$

if there exists a primitive matrix $P = P^{(r,n)}$, such that

$$a(S[P]) \neq 0,$$

and such that S is minimal with this property in the following sense: If S admits a representation

$$S = S_0[U]; \quad U, S_0 \text{ integral} \quad 0 < \det S_0 < \det S,$$

then for each primitive matrix P_0

$$a(S_0[P_0]) = 0.$$

Kernel forms always exist, if f is not identically 0. Of course the set of all f- kernel forms consists of finitely many unimodular classes.

4.4 Lemma. *Let S be a kernel form of a function $f \in \mathbf{P}$, and $Y = Y^{(r,n-r)}$ a rational matrix, such that $S[E, Y]$ is integral and such that $a(S[E, Y])$ is different from 0. Then Y is integral.*

Proof. Because the matrix $S[E, Y]$ is integral, positive semidefinite, and of rank r, it admits a representation

$$S[E, Y] = S_0[P],$$

where $S_0 = S_0^{(r)} > 0$ is integral and P is primitive. As a consequence S is representable by S_0. From the definition of a kernel form follows that both matrices are unimodular equivalent, we can assume that they are equal. Comparing both sides shows now that Y is integral.

In the case of a kernel form the hidden relations can be rewritten as

4.5 Proposition. *Let S be a kernel form of a function $f \in \mathbf{M}$. Then the relation*

$$\sum_{Y \in (\mathbb{Z}/q\mathbb{Z})^{(r, n-r)}} \alpha((E, Y)U) \exp \frac{2\pi i}{q} \sigma(Y'W) = 0$$

holds for all $U \in \mathrm{SL}(n, \mathbb{Z})$ and for all integral W such that $S^{-1}[W]$ is not integral.

4.6 Proposition. *Let S be a kernel form of a function $f \in \mathbf{M}$. Then $q^2 S^{-1}$ is*

  a) *integral if $n > r - 1$,*

  b) *half integral if $n = r - 1$.*

Proof. Our proof works only in the case $n > r + 1$:
We may assume that $a(S[E, 0]) \neq 0$. The inversion formula (4.3) for β instead of α shows the existence of an H, such that $b(S[E, H]) \neq 0$. Now the inversion formule gives us $\beta(E, H) = \beta(E, H + qS^{-1}X)$ for integral X. Hence $S[E, H + qS^{-1}X]$ is integral for integral X. It is easy to show and can be left to the reader that in case $n > r + 1$ the existence of such an H implies that $q^{-2}S$ is integral. (In case $n = r + 1$ this condition does not imply that $q^{-2}S$ is half integral.)

The case $n = r + 1$ is more involved. We leave it to the reader who may be interested in this border case.

5   A combinatorial problem

We want to discuss the following

5.1 Problem. *Let S be a kernel form of a function $f \in \mathbf{M}$. Does there exist a finite linear combination of theta series*

$$g(Z) = \sum_{\nu=1}^{h} \sum_{G \text{ integral}} \sum_{\mu=1}^{m_\nu} P_\nu(S^{1/2}G) \exp \frac{\pi i}{q} \sigma(S[G]Z + 2V'_{\nu\mu}G),$$

where P_ν are harmonic forms and $V_{\nu\mu}$ isotropic matrices with respect to S, such that

$$b(S[P]) = 0 \text{ for all primitive } P,$$

where $b(T)$ denote the Fourier coefficients of the difference $f - g$?

We assume for a moment that the problem has an affirmative answer. We apply it to a kernel form of f with minimal determinant. If $f - g$ is different from 0 we apply it to a kernel form with minimal determinant of $f - g$ and so on. The number of unimodular classes of kernel forms being finite, we obtain

If the above problem has an affirmative answer, the space $\mathbf{M}$ *mod* $\mathbf{P}(n, r - 1, q, \varrho_0)$ *is generated by the theta series described in 1.7.*

Especially the conjecture formulated in sec.1 would be true!

The "hidden relations" allow us to reduce the problem (for each fixed triple (n, r, q) to an elementary finite combinatorial problem:

We consider a basis $e_1, \ldots, e_h$ of the linear space generated by the vectors $\alpha(S[P])$, P primitive. We denote by α_ν the components of α with respect to this basis. For each vector e_ν there exists a harmonic form P_ν with the property (3.4)

$$P_\nu(S^{1/2}, 0) = E(S)^{-1} e_\nu.$$

What we need is a system of isotropic matrices $V_{\nu\mu}, 1 \le \mu \le m_\nu$, for each ν such that

$$a(S[P]) = \sum_{U \in \mathcal{E}(S)} \sum_{\nu=1}^{h} \sum_{\mu=1}^{m_\nu} P_\nu(S^{1/2}UP) \exp \frac{2\pi i}{q} \sigma(V'_{\nu\mu}UP).$$

($\mathcal{E}(S)$ denotes the unit group).
By definition of the polynomials P_ν, this equation is equivalent to

$$\alpha_\nu(S[P]) = E(S)^{-1} \sum_{U \in \mathcal{E}(S)} \sum_{\mu=1}^{m_\nu} \exp \frac{\pi i}{q} \sigma(V'_{\nu\mu}UP).$$

The harmonic polynomials have disappeared!
Because of the trivial invariance property $S[P] = S[UP]$ for $U \in \mathcal{E}(S)$, it is sufficient to prove the existence of a representation

$$\alpha_\nu(S[P]) = \sum_{\mu=1}^{m_\nu} \exp \frac{\pi i}{q} \sigma(V'_{\nu\mu}P)$$

with isotropic $V_{\nu\mu}$. Now our problem is reduced to what we call

The fundamental lemma. *Put*

$$T = \begin{cases} 2q^2 S^{-1} & \text{if } n = r + 1, \\ q^2 S^{-1} & \text{if } n > r + 1. \end{cases}$$

An integral matrix $H = H^{(r, n-r)}$ *is called isotropic, if*

$$T[H + qX] \equiv 0 \bmod 2q^2 \quad \text{for all integral } X.$$

An integral matrix $V = V^{(r,n)}$ *is called isotropic, if for each* $U \in Sl(n, \mathbb{Z})$ *the matrix* $H = H^{(r,n)}$ *defined by* $VU = (*, H)$ *is isotropic.*

We denote by $\mathcal{W}$ the vector space, which is generated by the functions

$$\alpha : \mathcal{P} \longrightarrow \mathbb{C},$$

which satisfy the "hidden relations"

$$\sum_{Y \in (\mathbb{Z}/q\mathbb{Z})^{(r,n-r)}} \alpha((E,Y)U) \exp \frac{2\pi i}{q} \sigma(Y'W)$$

for all unimodular U and all anisotropic $W = W^{(r,n-r)}$.

Is the space $\mathcal{W}$ generated by the special functions

$$P \longmapsto \exp \frac{2\pi i}{q} \sigma(P'V), \ V = V^{(r,n)} \ isotropic?$$

Of course this lemma depends only on T mod $2q^2$ and on V, W and U mod q. It hence is a finite problem for given (n, r, q).

Though it could not be proved in general, we prove it when $n \geq 2r$.

V   The fundamental lemma

1   Formulation of the lemma

In this section we propose a general variant of the "fundamental lemma" introduced in IV sec.5.

Let R be a finite commutative principal ideal ring with unit $1 = 1_R$. We assume $1 \neq 0$. For our purpose $R = \mathbb{Z}/q\mathbb{Z}$ would be sufficient. We assume that a character

$$e : R \longrightarrow \mathbb{C}^*$$
$$(e(x + y) = e(x)e(y)),$$

of the additive group of R is given. This character is assumed to be non-degenerate in the sense that every other character is of the form

$$r \longmapsto e(ar), \quad a \in R.$$

This means that the restriction of e to any ideal different from 0 is non-trivial.
Example:

$$R = \mathbb{Z}/q\mathbb{Z},$$
$$e(x) = e^{2\pi i x/q}.$$

We will be interested in finitely generated modules M over R and in their dual modules

$$M^* = \mathrm{Hom}_R(M, R).$$

By the theorem of elementary divisors each finitely generated module is a direct product of factor modules of R. The following properties of the dualizing functor are easy:

1) The functor

$$M \longmapsto M^*$$

is exact.

2) The modules M and M^* are (non-canonically) isomorphic; in particular they have the same order.

3) The canonical map

$$M \longrightarrow M^{**}$$
$$m \longmapsto \big(l \mapsto l(m)\big),$$

is an isomorphism.

To each element
$$l \in M^*$$
we associate a function
$$e_l : M \longrightarrow \mathbb{C},$$
$$e_l(m) = e(l(m)).$$

By the well-known orthogonality relations of characters they represent an orthogonal basis of the vector space of all functions $f : M \to \mathbb{C}$ with respect to the inner product

$$< f, g >= \sum_{m \in M} f(m)\overline{g(m)}.$$

1.1 Definition. *An isotropic structure on an R-module M is a set of –so-called isotropic– submodules*
$$L \subset M^* = \operatorname{Hom}_R(M, R)$$

of the dual module with the following properties:

1) The 0-module is isotropic.

2) Each submodule of an isotropic module is isotropic.

3) Assume that L_i, $1 \le i \le k$ is a system of isotropic submodules such that their annihilators
$$\mathbf{a}_i = \{m \in R, \quad m \cdot L_i = 0\}$$

are pairwise coprime $(\mathbf{a}_i + \mathbf{a}_j = R$ for $i \ne j)$. Then their sum $L = L_1 + \ldots + L_k$ is

isotropic.

1.2 Definition. *Let M be an R-module which has been equipped with an isotropic structure. Let furthermore m be a natural number. A vector of m elements of M^**

$$V = (v_1, \ldots, v_m) \in (M^*)^m = M^* \times \ldots \times M^*$$

is called isotropic, if the module

$$Rv_1 + \ldots + Rv_m \subset M^*$$

is isotropic.

Remark. The module $(M^*)^m$ and the dual module of M^m are canonically isomorphic:

$$(M^*)^m \times M^m \longrightarrow \mathbb{C},$$
$$(l_1, \ldots, l_m), (v_1, \ldots, v_m) \longmapsto l_1(v_1) + \ldots + l_m(v_m).$$

As a consequence, the function
$$e_V : M^m \longrightarrow \mathbb{C}$$
is well-defined for an element
$$V = (l_1, \ldots, l_m) \in (M^*)^m.$$

Sometimes we will use the
Notation:

$$e(V, P) = e_V(P) \quad \left(= e\left(\sum_j l_j(p_j)\right)\right), \quad V \in (M^*)^m, \; P \in M^m.$$

1.3 Definition. *Let*

$$\mathcal{M} \subset M^m$$

be any subset. A function

$$f : \mathcal{M} \longrightarrow \mathbb{C}$$

degenerates with respect to a given isotropic structure of M, if it can be written as linear combination of the functions

$$e_V|\mathcal{M}; \quad V \text{ isotropic.}$$

1.4 Remark. *In the case $\mathcal{M} = M^m$ the functions $e_V|\mathcal{M}$ are linearly independent. A function f degenerates if and only if it satisfies the following system of linear equations:*

$$\sum_{X \in M^m} f(X)e(X, W) = 0,$$

where W runs through all anisotropic (= not isotropic) vectors.

Besides $\mathcal{M} = M^m$, we will be interested in the set of primitive vectors:
A vector

$$P = (p_1 \dots p_m) \in M^m$$

is called **primitive**, if

$$M = Rp_1 + \dots + Rp_m,$$

or, equivalently, by the "Nakayama lemma", for each maximal ideal $\mathbf{m} \subset R$, the cosets of the elements p_i mod $\mathbf{m}M$ generate the vector space

$$M/\mathbf{m}M = M \otimes_R k, \quad k := R/\mathbf{m}.$$

We denote by $\mathcal{P}(M, m)$ the set of primitive vectors in M^m.

1.5 Definition. *The rank of M is*

$$\operatorname{rank} M := \max_{\mathbf{m}} \dim M/\mathbf{m}M.$$

The module M is called pure if

$$\operatorname{rank} M = \dim M/\mathbf{m}M \quad \text{for all maximal ideals } \mathbf{m}.$$

The set $\mathcal{P}(M, m)$ is not empty if and only if $m \geq \operatorname{rank} M$.

1.6 Remark. *A module of rank r can be generated by r elements.*

Proof. The ring R is the direct product of finitely many local rings. The remark is a direct consequence of the Nakayama lemma.

1.7 Definition. *Let M be a finitely generated pure R-module of rank r, which has been equipped with an isotropic structure. Let furthermore n be a natural number with the property*

$$n > r.$$

The pair (M, n) is called admissible, if the following two conditions on a function

$$f : \mathcal{P}(M, n) \longrightarrow \mathbb{C}$$

are equivalent:

1) f degenerates.

2) Let $E = (e_1, \ldots, e_r)$ be a (minimal) system of generators of the module M. The function

$$f_U : M^{n-r} \longrightarrow \mathbb{C}$$
$$f_U(Y) = f((E, Y)U)$$

degenerates for each $U \in \mathrm{GL}(n, R)$.

(The group $Gl(n, R)$ acts on

$$M^n = M^r \times M^{n-r}$$

in an obvious way.) Condition 2) is independent of the choice of a system of representatives.

Our problem is to find the conditions under which a pair (M, n) is admissible.

One of our main results will be that in the case $n \geq 2r$, the pair is always admissible. The following is an example of an isotropic structure which is not admissible in the case $r < n < 2r$.

$$L \subset M^* \text{ isotropic} \Longleftrightarrow L \neq M^*.$$

In our applications only very special isotropic structures occur. They satisfy the

1.8 Condition. *A module $L \subset M^*$ is isotropic, if each submodule of L of rank $\leq n - r$ is isotropic.*

In the case $n \geq 2r$ this condition is automatically satisfied!

1.9 Problem. *Assume that condition 1.8 is satisfied. Is each pair (M, n), $n > r$ admissible?*

In the following sections we shall give an affirmative answer in most cases. Some cases are still out-standing.

We recall the isotropic structures which occur in our theory of singular modular forms.

Let $q > 1$ by a natural number. We consider the ring $R = \mathbb{Z}/q\mathbb{Z}$ and the free module

$$M = (\mathbb{Z}/q\mathbb{Z})^r,$$

which is pure of rank r. The elements of M^m can be considered as $r \times m$ matrices. For $m \geq r$, such a matrix P corresponds to a primitive element, if P is part of an invertible $m \times m$-matrix

$$U = \begin{pmatrix} P \\ * \end{pmatrix} \in \mathrm{GL}(m, R).$$

We identify M with its dual module M^* by means of

$$M \longrightarrow M^*$$
$$A \longmapsto \left(X \mapsto \sigma(A' \cdot X) \right).$$

Let a natural number n, $n > r$ be given. For the definition of the isotropic structure, we need a symmetric integral $r \times r$ matrix S which is invertible and is such that

$$q^2 S^{-1} \text{ is } \begin{cases} \text{integral} & \text{if } n > r + 1 \\ \text{half integral} & \text{if } n = r + 1. \end{cases}$$

1.10 Definition. *A subgroup $L \subset R^r$ is called isotropic (with respect to S), if for each vector $v \in Z^r$, whose coset $\bmod q$ is contained in L,*

$$S^{-1}[v] \quad and \begin{cases} S^{-1}v & \text{if } n > r + 1 \\ 2S^{-1}v & \text{if } n = r + 1 \end{cases}$$

are integral.

It is easy to see that the axioms 1.1 are satisfied.

1.11 Remark. *The coset of a matrix $V \in Z^{(r,m)} \bmod q$ (considered as an element of M^m) is isotropic in sense of 1.2 with respect to the isotropic structure 1.10 if and only if*

$$S^{-1}[V] \quad and \begin{cases} S^{-1}V & \text{if } n > r + 1 \\ 2S^{-1}V & \text{if } n = r + 1 \end{cases}$$

are integral.

This is consistent with the definition of isotropy, which we introduced in IV sec.5.

For our purpose only the isotropic structures 1.10 are of interest. Nevertheless the question arises, what exactly is needed from the condition of "isotropic structure". This is one reason, why we proposed the abstract formulation. Another reason is that during the proof of the fundemental lemma (which is complete in the case $n \geq 2r$), even in the case 1.10, isotropic structures will occur which definitely are not of the type 1.10. For example we will have to consider modules M which are not free.

2   Reduction to the case of a local ring R

As we already mentioned, a finite commutative principal ideal ring is the direct product of local rings,

$$R = \prod_{i=1}^{k} R_i.$$

In the case of our interest $\mathbb{Z}/q\mathbb{Z}$, this decomposition comes from the prime decomposition of q,

$$q = \prod_{i=1}^{k} p_i^{\nu_i},$$

where $p_1, \ldots, p_k$ are pairwise distinct primes, namely

$$\mathbb{Z}/q\mathbb{Z} = \prod_{i=1}^{k} \mathbb{Z}/p^{\nu_i}\mathbb{Z}.$$

We want to reduce the fundamental lemma to the local factors of R. Let M be a finitely generated R-module. We consider R_i as R-module with respect to the natural projection $R \to R_i$. We define

$$M_i = M \otimes R_i,$$

which is a R_i- (and also an R-) module. If L is any submodule of M then $L_i := L \otimes R_i$ can be considered as a sub-module of M_i.

2.1 Remark. *There is a one-to-one correspondence between the submodules L of M and systems of submodules L_i of M_i $(1 \leq i \leq k)$.*

An immediate consequence of the definition of an isotropic structure, especially the assumption 3) in 1.1 is

2.2 Remark. *There is a one-to-one-correspondence between isotropic structures on M and systems of isotropic structures on M_i $(1 \leq i \leq k)$.*
A vector $V = (v_1, \ldots v_m) \in (M^*)^m$ is isotropic, if and only if all the k components $V_i \in (M_i^*)^m$ are isotropic.

A similar correspondence holds for the set of primitive vectors. In the following we assume that M is pure, i.e. that all components M_i are of the same rank r.

2.3 Remark. *The natural projection defines a bijection*

$$\mathcal{P}(M, m) \longrightarrow \prod_{i=1}^{k} \mathcal{P}(M_i, m).$$

We are able to prove the announced reduction to the local case.

2.4 Lemma. *Let n be a natural number such that $n > r$. Assume that all (M_i, n) $(1 \leq i \leq k)$ are admissible (1.7). Then (M, n) is admissible.*

Proof. Let $E = (e_1, \ldots, e_r)$ be a minimal system of generators of M. We have to consider functions

$$f : \mathcal{P}(M, n) \longrightarrow \mathbb{C}.$$

We assume that

$$Y \longmapsto f((E, Y)U)$$

degenerates for all $U \in \mathrm{GL}(r, R)$. We consider $f(P_1, \ldots, P_k)$ as function on

$$\prod_{i=1}^{k} \mathcal{P}(M_i, n).$$

The above statement carries over to each function

$$P_i \longmapsto f(P_1, \ldots, P_k),$$

if $(P_1, \ldots P_{i-1}, P_{i+1} \ldots P_k)$ is fixed. Therefore this function degenerates, because (M_i, n) is admissible. Hence it is linear combination of functions

$$P_i \longmapsto e(V_i, P_i); \quad V_i \text{ isotropic.}$$

The following lemma shows that f is linear combination of functions

$$e(V, P) = \prod_{i=1}^{k} e(V_i, P_i), \quad V \text{ isotropic.}$$

2.5 Lemma. *Let $\mathcal{P}_1$, $\mathcal{P}_2$ be two sets and*

$$\mathcal{W}_\nu \subset \mathbb{C}^{\mathcal{P}_\nu} \quad (\nu = 1, 2)$$

two finite dimensional $\mathbb{C}$-vector spaces of complex valued functions on $\mathcal{P}_\nu$.
The following two conditions for a function are equivalent

$$f : \mathcal{P}_1 \times \mathcal{P}_2 \longrightarrow \mathbb{C}.$$

1) $f(P_1, P_2)$ *is contained in $\mathcal{W}_1$ for fixed P_2 and conversely.*
2) f *is a finite linear combination of functions*

$$f_1(P_1) \cdot f_2(P_2), \quad f_\nu \in \mathcal{W}_\nu \ (\nu = 1, 2).$$

Proof. The direction "2) $\Longrightarrow$ 1)" is trivial

1) $\Longrightarrow$ 2):   We choose bases

$$g_1, \ldots, g_k \text{ of } \mathcal{W}_1,$$
$$h_1, \ldots, h_l \text{ of } \mathcal{W}_2.$$

We have

$$f(P_1, P_2) = \sum_j C_j(P_1) \cdot h_j(P_2)$$

for fixed P_1. We have to show that the functions C_j are contained in $\mathcal{W}_1$. For this purpose we choose points

$$P_2^{(1)}, \ldots P_2^{(l)} \in \mathcal{P}_2,$$

such that the matrix

$$\left(h_j(P_2^{(i)}) \right)_{1 \leq i, j \leq l}$$

is invertible. The functions C_j are then linear combinations of the functions

$$P_1 \longmapsto f(P_1, P_2^{(\nu)}); \quad 1 \leq \nu \leq l.$$

3   The case, where the sum of isotropic vectors is isotropic

We consider a finite principal ideal ring R and a finitely generated R-module M of rank r, which is equipped with an isotropic structure.

3.1 Proposition.   *Assume that there exists a submodule $L \subset M^*$, such that the isotropic modules are precisely the submodules of L. Then any pair (M, n), $n > r$, is admissible.*

Proof. We denote by $L^\perp$ the orthogonal complement of L in M:

$$L^\perp := \{ x \in M, \ l(x) = 0 \text{ for all } l \in L \}.$$

A vector $V = (v_1, \ldots, v_m) \in (M^*)^m$ (m some natural number) is isotropic if and only if its components v_i are isotropic, i.e. contained in L. By assumption the set of isotropic vectors is a group. It follows from the well-known orthogonality relation of characters that for a function

$$f : \mathcal{M} \longrightarrow \mathbb{C}, \quad \mathcal{M} \subset M^m,$$

on a subset $\mathcal{M}$ of M^m, the following two conditions are equivalent:

a) f is a linear combination of functions $e(V, Y)$, V isotropic,

b) $f(Y + H) = f(Y)$ for $H, Y + H \in \mathcal{M}$ and for $H \in (L^\perp)^m$.

The conclusion b) $\Rightarrow$ a) follows from the fact that f extends to a periodic function on the whole space M^m (but such an extension is not unique).

The fundamental lemma in this special case is the following:

Let $f : \mathcal{P}(M, n) \to \mathbb{C}$ be a function such that

$$Y \longmapsto f((E, Y)U)$$

is periodic with respect to $(L^\perp)^{n-r}$. Then f is periodic with respect to $(L^\perp)^n$.
There is no loss of generality, if we make the following assumptions.

1) $L = 0$ (replacing M by M/L).
2) $R = \mathbb{Z}/q\mathbb{Z}$ (replacing R by the subring generated by 1_r).
3) $q = p^m$, p a prime (applying 2.4).
4) M is a free module (replacing M by a free module, which has M as homomorphic image).

We reformulate the statement:

Let $\mathcal{P}$ be the set of primitive $r \times n$-matrices with coefficients in $\mathbb{Z}/p^m\mathbb{Z}$. For each matrix $U \in \mathrm{GL}(n, \mathbb{Z}/p^m\mathbb{Z})$ we consider the subset

$$\mathcal{B}(U) = \{P = (E, Y)U'^{-1},\ Y \in (\mathbb{Z}/p^m\mathbb{Z})^{n-r}\},$$

where E denotes the $r \times r$-unit matrix. Our statement is:

Let $f : \mathcal{P} \to \mathbb{C}$ be a function such that the restriction of f to each $\mathcal{B}(U)$ is constant. Then f is constant.

Applying the assumption about f to

$$U'^{-1} = \begin{pmatrix} A & 0 \\ 0 & E \end{pmatrix}, \quad A \in \mathrm{GL}(n, R),$$

we obtain:

For each $A \in \mathrm{GL}(n, \mathbb{Z})$ the function $Y \mapsto f(A, Y)$ is constant.
We introduce

$$F : \mathrm{GL}(r, R) \longrightarrow \mathbb{C},$$
$$F(A) = f(A, E).$$

The next lemma shows that it is enough to prove that F is constant.

3.2 Lemma. *Each set $\mathcal{B}(U)$ contains an element $(A, *)$, $A \in \mathrm{GL}(n, R)$.*

Proof. We write

$$U'^{-1} = \begin{pmatrix} A & B \\ C & D \end{pmatrix}, \quad A \in R^{(r,r)}.$$

Because of

$$(E, Y)U'^{-1} = (A + YC, *)$$

it is sufficient to construct a matrix Y such that $\det(A + YC)$ is invertible. An element in R is invertible if and only if its image in $\mathbb{Z}/p\mathbb{Z}$ is different from 0. Hence we can assume that R is a field. The statement is invariant with respect to a substitution of the type

$$A \mapsto U_1 A U_2, \quad Y \mapsto U_1 Y, \quad C \mapsto C U_2,$$

where U_1, U_2 are invertible matrices. Therefore we may assume

$$A = \begin{pmatrix} E^{(s)} & 0 \\ 0 & 0 \end{pmatrix}, \quad E^{(s)} \ s \times s\text{-unit matrix}.$$

We decompose C

$$C = (C_1, C_2); \quad C_1 \in R^{(n-r,s)}, \ C_2 \in R^{(n-r,r-s)}.$$

The matrix C_2 has rank $n - r$. Therefore we have $r - s \geq n - r$. We choose Y as

$$Y = \begin{pmatrix} 0 \\ Y_2 \end{pmatrix}; \quad Y_2 \in R^{(r-s,n-r)}.$$

The condition for Y_2 is

$$\det(Y_2 C_2) \neq 0.$$

We may assume

$$C_2 = (E, 0)$$

and can take

$$Y_2 = \begin{pmatrix} E \\ 0 \end{pmatrix}.$$

Lemma 3.2 is proved.

We make use of the equation

$$f((A, 0) \cdot U'^{-1}) = f((A, Y)\dot{U}'^{-1})$$

in the special case

$$U'^{-1} = \begin{pmatrix} E & 0 \\ X & E \end{pmatrix}, \quad X \in R^{(n-r,r)}$$

and obtain:

Assume that $A, B \in \mathrm{GL}(r, R)$ are two invertible matrices such that

$$B - A = YX, \quad Y \in R^{(r,n-r)}, \ X \in R^{(n-r,r)},$$

then

$$F(A) = F(B).$$

This condition for A, B is satisfied, if

$$\mathrm{rank}(A - B) \leq 1.$$

To complete the proof of 3.1, we show

3.3 Lemma. *Assume that $A, B \in \mathrm{GL}(r, R)$ are two invertible matrices with coefficients in a field R. Then there exists a finite system*

$$A = A_1, \ldots, A_s = B$$

of invertible matrices such that

$$\mathrm{rank}(A_{i+1} - A_i) \leq 1, \quad 1 \leq i < s.$$

Proof. We may assume that one of the two matrices is the unit matrix and that the other has Jordan normal form, i.e. $B = E$, and

$$A = \begin{pmatrix} 0 & \cdot & \cdot & \cdot & 0 & a_1 \\ 1 & \cdot & & & & \cdot & \cdot \\ 0 & \cdot & & & & \cdot & \cdot \\ \cdot & \cdot & \cdot & \cdot & \cdot & \cdot \\ \cdot & & \cdot & \cdot & 0 & \cdot \\ 0 & \cdot & \cdot & 0 & 1 & a_r \end{pmatrix}.$$

We may replace A by

$$A = \begin{pmatrix} 0 & \cdot & \cdot & \cdot & 0 & 1 \\ 1 & \cdot & & & & \cdot & 0 \\ 0 & \cdot & & & & \cdot & \cdot \\ \cdot & \cdot & \cdot & \cdot & \cdot & \cdot \\ \cdot & & \cdot & \cdot & 0 & 0 \\ 0 & \cdot & \cdot & 0 & 1 & 0 \end{pmatrix}$$

because the difference between the two matrices is of rank ≤ 1. For this concrete matrix it is easy to find a connecting system, for example in the case $r = 3$

$$\begin{pmatrix} 0 & 0 & 1 \\ 1 & 0 & 0 \\ 0 & 1 & 0 \end{pmatrix} \quad \begin{pmatrix} 1 & 0 & 1 \\ 1 & 0 & 0 \\ 0 & 1 & 0 \end{pmatrix} \quad \begin{pmatrix} 1 & 0 & 1 \\ 1 & 1 & 0 \\ 0 & 1 & 0 \end{pmatrix} \quad \begin{pmatrix} 1 & 0 & 1 \\ 1 & 1 & 0 \\ 1 & 1 & 1 \end{pmatrix}$$

$$\begin{pmatrix} 1 & 0 & 0 \\ 1 & 1 & 0 \\ 1 & 1 & 1 \end{pmatrix} \quad \begin{pmatrix} 1 & 0 & 0 \\ 0 & 1 & 0 \\ 0 & 1 & 1 \end{pmatrix} \quad \begin{pmatrix} 1 & 0 & 0 \\ 0 & 1 & 0 \\ 0 & 0 & 1 \end{pmatrix}.$$

All the matrices occuring here are invertible! Now the proof of 3.1 is complete.

The case that there exists a biggest isotropic module is very exceptional. But it occurs in the theory of modular forms, namely in the case $r = 1$. Let s be a natural number, which we consider as quadratic form in one variable. Let q be a level such that q^2/s is integral. The associate isotropic structure consists of all $v \in \mathbb{Z}/q\mathbb{Z}$, such that

$$v^2/s \quad \text{and} \quad \begin{cases} 2v/s & \text{if } n = 2 \\ v/s & \text{if } n > 2 \end{cases}$$

both are integral. The set of those v is an additive group. What we obtain is the following theorem:

Each modular form of weight 1/2 is a linear combination of theta series.

A more precice formulation will be given in VI sec.1. In the scalar valued case, this result is due to ENDRES [En].

4   The case $r = 2$

We investigate a special isotropic structure of rank 2, which will give the representation of modular forms of weight 1 as linear combination of theta series at least for square free levels q.

Let $k = \mathbb{Z}/p\mathbb{Z}$ be a finite prime field.

4.1 Definition. *A subspace $L \subset k^2$ is called isotropic if it is contained in one of the two coordinate axes.*

Corollary. *A matrix $H = \begin{pmatrix} a \\ b \end{pmatrix}$, $a, b \in k^{(1,m)}$ is isotropic if and only if $a = 0$ or $b = 0$.*

4.2 Proposition. *The pair (k^2, n) (equipped with the isotropic structure 4.1) is admissible for each $n > 2$.*

Proof. We denote by $\mathcal{P}$ the set of all primitive $2 \times n$-matrices. A subset $\mathcal{B} \subset \mathcal{P}$ is called an **orbit**, if there exists a matrix $U \in \mathrm{GL}(n, k)$, such that

$$\mathcal{B} = \mathcal{B}(U) := \left\{ (E, Y)U'^{-1}, \ Y \in k^{(2, n-2)} \right\}.$$

It is important to keep in mind that the orbits carry a vector space structure, i.e. the map

$$k^{(2, n-2)} \longrightarrow \mathcal{B}(U),$$
$$Y \longmapsto (E, Y)U'^{-1},$$

is a bijection. We have to investigate the space $\mathcal{W}$ of functions

$$f : \mathcal{P} \longrightarrow \mathbb{C},$$

such that for each $U \in \mathrm{GL}(n, k)$ the function

$$f_U : k^{(2, n-2)} \longrightarrow \mathbb{C},$$
$$f_U(Y) = f((E, Y)U'^{-1}),$$

is a linear combination of the characters

$$Y \longmapsto e(V, Y), \quad V \text{ isotropic in } k^{(2, n-2)}.$$

Proposition 4.2 states that each function $f \in \mathcal{W}$ is linear combination of functions

$$P \longmapsto e(V, P), \quad V \text{ isotropic in } k^{(2, n)}.$$

4.3 Lemma. *For a function*

$$B : k^{(2, m)} \longrightarrow \mathbb{C}$$

the following three statements are equivalent:

1) B is linear combination of characters $Y \mapsto e(V, Y)$, V isotropic.

2) *There exist two functions*

$$B_\nu : k^{(1,m)} \longrightarrow \mathbb{C}, \quad \nu = 1, 2,$$

such that

$$B \begin{pmatrix} a \\ b \end{pmatrix} = B_1(a) + B_2(b).$$

3) *The system of linear equations*

$$B \begin{pmatrix} a \\ b \end{pmatrix} + B \begin{pmatrix} c \\ d \end{pmatrix} = B \begin{pmatrix} a \\ d \end{pmatrix} + B \begin{pmatrix} c \\ b \end{pmatrix}.$$

is satisfied for all vectors $a, b, c, d \in k^{(1,m)}$.

Corollary. *A function*

$$A : \mathcal{P} \longrightarrow \mathbb{C}$$

is contained in the space $\mathcal{W}$ if and only if for

$$P = \begin{pmatrix} a \\ b \end{pmatrix}, \quad Q = \begin{pmatrix} c \\ d \end{pmatrix} \in \mathcal{P}$$

$$(a, b, c, d \in k^{(1,m)})$$

any two primitive matrices, which are contained in a common orbit

$$A \begin{pmatrix} a \\ b \end{pmatrix} + A \begin{pmatrix} c \\ d \end{pmatrix} = A \begin{pmatrix} a \\ d \end{pmatrix} + A \begin{pmatrix} c \\ b \end{pmatrix}.$$

The proof is easy and can be left to the reader. A similar statement is true for the set $\mathcal{P}$ of primitive matrices instead of $k^{(2,n)}$.

4.4 Lemma. *For a function*

$$A : \mathcal{P} \longrightarrow \mathbb{C},$$

the following three statements are equivalent:

1) *A is a linear combination of characters $P \mapsto e(V, P)$, V isotropic.*
2) *There exist two functions*

$$A_\nu : k^{(1,n)} - \{0\} \longrightarrow \mathbb{C}, \quad \nu = 1, 2,$$

such that

$$A \begin{pmatrix} a \\ b \end{pmatrix} = A_1(a) + A_2(b)$$

whenever the matrix $\begin{pmatrix} a \\ b \end{pmatrix}$ is primitive.

3) *The system of linear equations*

$$A \begin{pmatrix} a \\ b \end{pmatrix} + A \begin{pmatrix} c \\ d \end{pmatrix} = A \begin{pmatrix} a \\ d \end{pmatrix} + A \begin{pmatrix} c \\ b \end{pmatrix}$$

is satisfied for all quadruples $a, b, c, d \in k^{(1,n)}$, such that the matrices

$$\begin{pmatrix} a \\ b \end{pmatrix}, \quad \begin{pmatrix} c \\ d \end{pmatrix}, \quad \begin{pmatrix} a \\ d \end{pmatrix}, \quad \begin{pmatrix} c \\ b \end{pmatrix}$$

are primitive.

Corollary. *The dimension of the space of functions A is*

$$2p^n - 3.$$

Proof. We show the only non-trivial direction 3) $\rightarrow$ 2):

We choose a vector $e \in k^n$, which is different from 0. In a first step we want to show that the function

$$\begin{pmatrix} a \\ b \end{pmatrix} \longmapsto A\begin{pmatrix} a \\ b \end{pmatrix} - A\begin{pmatrix} a \\ e \end{pmatrix} - A\begin{pmatrix} b \\ e \end{pmatrix}$$

is constant. It is defined on the set

$$\mathcal{P}_0 := \left\{ \begin{pmatrix} a \\ b \end{pmatrix} \in \mathcal{P}; \quad a, b \notin ke \right\}.$$

Claim. Two elements

$$\begin{pmatrix} a \\ b \end{pmatrix}, \quad \begin{pmatrix} c \\ d \end{pmatrix} \in \mathcal{P}_0$$

can be connected by a chain

$$\begin{pmatrix} a \\ b \end{pmatrix} = \begin{pmatrix} a_0 \\ b_0 \end{pmatrix}, \begin{pmatrix} a_1 \\ b_1 \end{pmatrix}, \dots, \begin{pmatrix} a_m \\ b_m \end{pmatrix} = \begin{pmatrix} c \\ d \end{pmatrix}$$

of matrices in $\mathcal{P}_0$ such that for each $\nu \in \{1, \dots, m\}$

$$\text{either} \quad a_\nu = a_{\nu-1} \quad \text{or} \quad b_\nu = b_{\nu-1}.$$

Proof of the claim. If a, b, c, d are linearly independent, we can take the connection

$$\begin{pmatrix} a \\ b \end{pmatrix}, \begin{pmatrix} a \\ d \end{pmatrix}, \begin{pmatrix} c \\ d \end{pmatrix}.$$

Therefore assume that they are dependent. Then one can find a vector c', such that

$$c' \notin ke, \quad c' \notin ka + kb + kc + kd.$$

The existence of c' follows from the inequality $p^n > p + p^{n-1} - 1$. We can take the connection

$$\begin{pmatrix} a \\ b \end{pmatrix}, \begin{pmatrix} a \\ c' \end{pmatrix}, \begin{pmatrix} c \\ c' \end{pmatrix}, \begin{pmatrix} c \\ d \end{pmatrix}.$$

Because of the claim, the function

$$\begin{pmatrix} a \\ b \end{pmatrix} \longmapsto A\begin{pmatrix} a \\ b \end{pmatrix} - A\begin{pmatrix} a \\ e \end{pmatrix} - A\begin{pmatrix} b \\ e \end{pmatrix}$$

will be constant, if it is constant as function of a for fixed b and conversely. We prove the first statement, i.e.

$$A\begin{pmatrix} a \\ b \end{pmatrix} - A\begin{pmatrix} a \\ e \end{pmatrix} - A\begin{pmatrix} b \\ e \end{pmatrix} = A\begin{pmatrix} a' \\ b \end{pmatrix} - A\begin{pmatrix} a' \\ e \end{pmatrix} - A\begin{pmatrix} b \\ e \end{pmatrix}$$

or

$$A\begin{pmatrix} a \\ b \end{pmatrix} + A\begin{pmatrix} a' \\ e \end{pmatrix} = A\begin{pmatrix} a \\ e \end{pmatrix} + A\begin{pmatrix} a' \\ b \end{pmatrix}.$$

But this is one of the relations 4.4 3).

To complete the proof of 4.4, we define functions A_1 and A_2. For this purpose, we choose a vector $f \in k^n$, which is independent of e. Then we define

$$A_1(e) = 0,$$

$$A_2(e) = -A\begin{pmatrix} a \\ b \end{pmatrix} + A\begin{pmatrix} a \\ e \end{pmatrix} + A\begin{pmatrix} e \\ b \end{pmatrix}, \qquad \begin{pmatrix} a \\ b \end{pmatrix} \in \mathcal{P}_0,$$

$$A_1(a) = A\begin{pmatrix} a \\ e \end{pmatrix} - A_2(a), \qquad\qquad\qquad \text{if } a \notin ke,$$

$$A_2(b) = A\begin{pmatrix} e \\ b \end{pmatrix}, \qquad\qquad\qquad\qquad \text{if } b \notin ke,$$

$$A_1(te) = A\begin{pmatrix} te \\ f \end{pmatrix} - A_2(f), \qquad\qquad \text{for } t \neq 0, 1,$$

$$A_2(te) = A\begin{pmatrix} f \\ te \end{pmatrix} - A_1(f), \qquad\qquad \text{for } t \neq 0, 1.$$

We have to prove the relation

$$A\begin{pmatrix} a \\ b \end{pmatrix} = A_1(a) + A_2(b).$$

By construction this relation is true for matrices in $\mathcal{P}_0$. We have to prove it in the case when a or b is dependent of e.

1) a is dependent of e. Then b and e are independent. We have to prove

$$A\begin{pmatrix} te \\ b \end{pmatrix} = A_1(te) + A\begin{pmatrix} e \\ b \end{pmatrix}.$$

If $t = 1$, this relation holds by definition. In the case $t \neq 1$ the statement is

$$A\begin{pmatrix} te \\ b \end{pmatrix} = A_1(te) + A\begin{pmatrix} te \\ f \end{pmatrix} - A\begin{pmatrix} e \\ f \end{pmatrix} = A_1(te) + A\begin{pmatrix} e \\ b \end{pmatrix},$$

which follows from the assumption about A.

1) b is dependent of e. Then a and e are independent. The statement is

$$A\begin{pmatrix} a \\ te \end{pmatrix} = A\begin{pmatrix} a \\ e \end{pmatrix} - A_2(e) + A_2(te).$$

Again this statement is trivial in the case $t = 1$ and for $t \neq 1$ it is a consequence of the assumptions about A.

4.5 Remark. *Let a, b, c, d be four linearly independent vectors from $k^{(1,n)}$. The two matrices*

$$P = \begin{pmatrix} a \\ b \end{pmatrix}, \quad Q = \begin{pmatrix} c \\ d \end{pmatrix}$$

are contained in a common orbit.

Proof. The group $\mathrm{GL}(n, k)$ acts on the set of orbits by multiplication from the right. Hence we may assume

$$a = (1, 0, \ldots, 0); \quad b = (0, 1, 0, \ldots, 0);$$
$$c = (1, 0, 1, 0, \ldots, 0); \quad d = (0, 1, 0, 1, 0 \ldots, 0),$$

because these four vectors are linearly independent. But then the two matrices are contained in the standard orbit.

Next we want to reduce proposition 4.2 to the case $n = 3$ and assume that in this case it already has been proved. We have to prove the relation 4.4 3) under the assumption that it holds inside orbits

1. case. The vectors a, b, c, d are linearly independent. The statement follows from 4.5

2. case. The 4 vectors are linearly dependent. Then they are contained in a three-dimensional subspace $\mathcal{V}$. Wew may assume that this space is defined by the equations $v_4 = \ldots = v_n = 0$ and identify it with k^3. Now the reduction to the case $n = 3$ is obvious.

For the rest of the proof we may assume $n = 3$. Because of the corollary to 4.4, it is sufficient to show

$$\dim \mathcal{W} \leq 2p^3.$$

This follows from

4.6 Lemma. *Let e, f, g be any basis of k^3. A function $A \in \mathcal{W}$ vanishes identically, if it vanishes on all primitive matrices of the following type.*

$$\begin{pmatrix} a \\ e \end{pmatrix}, \quad \begin{pmatrix} e \\ b \end{pmatrix}, \quad \begin{pmatrix} te \\ f \end{pmatrix}, \quad \begin{pmatrix} f \\ te \end{pmatrix}, \quad \begin{pmatrix} f \\ g \end{pmatrix}$$
$$(t \in k - \{0\}).$$

Their number is

$$(p^3 - p) + (p^3 - p) + (p - 2) + (p - 2) + 1 = 2p^3 - 3.$$

We have to make use of the relations, which define $\mathcal{W}$, i.e. we have to construct pairs of primitive matrices, which are contained in a common orbit.

4.7 Lemma. *Let a, b, c be three linearly independent vectors from $k^{(1,3)}$ and t an arbitrary element of k. The two matrices*

$$P = \begin{pmatrix} a \\ b \end{pmatrix}, \quad Q = \begin{pmatrix} c \\ d \end{pmatrix}, \quad d = b + t(a - c),$$

are primitive and contained in a common orbit.

Proof. Again we make use of the fact that the group $GL(3, k)$ acts on the set of orbits by multiplication from the right. Hence we may assume

$$a = (1, 0, 0); \quad b = (0, 1, 0); \quad c = (1, 0, 1).$$

Then we have $d = (0, 1, *)$ and the two matrices are contained in the standard orbit.

We have to consider a certain subset $\mathcal{P}(1)$ of the set of primitive 2×3-matrices consisting of all matrices of the form

$$P = \begin{pmatrix} a \\ b \end{pmatrix}$$

with the property

$$a = x_1 f + x_2 g + x_3 e,$$
$$b = y_1 f + y_2 g + y_3 e,$$

where

$$x_1 y_2 - x_2 y_1 = 1.$$

Then the vectors a, b, e are linearly independent. From Lemma 4.7, it follows that the two matrices

$$P = \begin{pmatrix} a \\ b \end{pmatrix}; \quad Q = \begin{pmatrix} e \\ a + b - e \end{pmatrix}$$

are contained in a common orbit. The same is true for

$$P = \begin{pmatrix} a \\ b \end{pmatrix}; \quad Q = \begin{pmatrix} a + b - e \\ e \end{pmatrix}.$$

We obtain

4.8 Remark. *Let $A \in \mathcal{W}$ be a function with the property*

$$A \begin{pmatrix} a \\ e \end{pmatrix} = A \begin{pmatrix} e \\ a \end{pmatrix} = 0$$

(for all a which are independent of e). Then

$$A \begin{pmatrix} a \\ b \end{pmatrix} = A \begin{pmatrix} a \\ a + b - e \end{pmatrix} = A \begin{pmatrix} a + b - e \\ b \end{pmatrix}$$

for all

$$P = \begin{pmatrix} a \\ b \end{pmatrix} \in \mathcal{P}(1).$$

We are going to show that this property implies that A is constant on $\mathcal{P}(1)$. More precisely:

4.9 Lemma. *Let*

$$A : \mathcal{P}(1) \longrightarrow \mathbb{C}$$

be a function with the following two properties:

1) $A\begin{pmatrix} a \\ b \end{pmatrix} = A\begin{pmatrix} a \\ a+b-e \end{pmatrix} = A\begin{pmatrix} a+b-e \\ b \end{pmatrix}$ *for* $\begin{pmatrix} a \\ b \end{pmatrix} \in \mathcal{P}(1)$.

2) $A\begin{pmatrix} a \\ b \end{pmatrix} + A\begin{pmatrix} c \\ d \end{pmatrix} = A\begin{pmatrix} a \\ d \end{pmatrix} + A\begin{pmatrix} c \\ b \end{pmatrix}$ *if*

   (a) a, b, c are linearly independent,

   (b) $d = b + t(a - c)$ for a $t \in k$,

   (c) $\begin{pmatrix} a \\ b \end{pmatrix}, \begin{pmatrix} c \\ d \end{pmatrix}, \begin{pmatrix} a \\ d \end{pmatrix}, \begin{pmatrix} c \\ d \end{pmatrix}$ are contained in $\mathcal{P}$.

Then the function A is constant.

Proof. We define an action of the group $\mathrm{SL}(2, k)$ on $\mathcal{P}(1)$,

$$\mathrm{SL}(2, k) \times \mathcal{P}(1) \longrightarrow \mathcal{P}(1),$$
$$(U, P) \longmapsto U\{P\}$$

by means of the formulae

$$U\left\{ \begin{matrix} x_1 f + x_2 g + x_3 e \\ y_1 f + y_2 g + y_3 e \end{matrix} \right\} = \begin{pmatrix} x_1' f + x_2' g + x_3' e \\ y_1' f + y_2' g + y_3' e \end{pmatrix},$$

where

$$\begin{pmatrix} x_1' & x_2' & x_3' \\ y_1' & y_2' & y_3' \end{pmatrix} = U \cdot \begin{pmatrix} x_1 & x_2 & x_3 \\ y_1 & y_2 & y_3 \end{pmatrix} + \begin{pmatrix} 0 & 0 & 1 - u_{11} - u_{12} \\ 0 & 0 & 1 - u_{21} - u_{22} \end{pmatrix}.$$

The following formulae are easily verified:

(a) $E\{P\} = P,$

(b) $(UV)\{P\} = U\{V\{P\}\},$

(c) $\begin{pmatrix} 1 & 1 \\ 0 & 1 \end{pmatrix} \left\{ \begin{matrix} a \\ b \end{matrix} \right\} = \begin{pmatrix} a+b-e \\ b \end{pmatrix},$

             $\begin{pmatrix} 1 & 0 \\ 1 & 1 \end{pmatrix} \left\{ \begin{matrix} a \\ b \end{matrix} \right\} = \begin{pmatrix} a \\ a+b-e \end{pmatrix}.$

We consider the equivalence relation, which is generated by the relations

$$\begin{pmatrix} a \\ b \end{pmatrix} \sim \begin{pmatrix} a \\ a+b-e \end{pmatrix} \sim \begin{pmatrix} a+b-e \\ b \end{pmatrix}.$$

The group $\mathrm{SL}(2, k)$ is generated by the two matrices

$$\begin{pmatrix} 1 & 1 \\ 0 & 1 \end{pmatrix} \text{ and } \begin{pmatrix} 1 & 0 \\ 1 & 1 \end{pmatrix}.$$

From the definition of the action of $\mathrm{SL}(2, k)$ we obtain

4.10 Remark. *Each equivalence class contains a unique representative of the form*

$$\begin{pmatrix} f + \alpha e \\ g + \beta e \end{pmatrix}; \quad \alpha, \beta \in k.$$

By assumption the function A is constant on each equivalence class. For the proof of Lemma 4.9 it is therefore sufficient to show:

The function

$$B(\alpha, \beta) := A \begin{pmatrix} f + \alpha e \\ g + \beta e \end{pmatrix}$$

is constant on $k \times k$.

An immediate consequence of the definition of the action of $\mathrm{SL}(2, k)$ is

4.11 Remark. *We have*

$$A \begin{pmatrix} x_1 f + x_2 g + x_3 e \\ y_1 f + y_2 g + y_3 e \end{pmatrix} =$$
$$B(y_2 x_3 - x_2 y_3 + 1 - y_2 + x_2, y_3 x_1 - x_3 y - 1 + 1 + y_1 - x_1).$$

We have to make use of the conditions 4.9, 2).

Notation. *A quadruple* (a, b, c, d) *is called* **admissible,** *if the conditions 2)(a),(b),(c) in 4.9 are satisfied.*

For example the quadruple

$$(f, g, f + \alpha e, g + \beta e), \quad \alpha \neq 0,$$

is admissible. We obtain the relation

$$B(\alpha, \beta) = B(\alpha, 0) + B(0, \beta) - B(0, 0),$$

which is also true for $\alpha = 0$.

Another admissible quadruple is

$$(f, f + g, f + e, f + g + te) \quad (t \in k).$$

We obtain

$$A \begin{pmatrix} f \\ f + g \end{pmatrix} + A \begin{pmatrix} f + e \\ f + g + te \end{pmatrix} = A \begin{pmatrix} f \\ f + g + te \end{pmatrix} + A \begin{pmatrix} f + e \\ f + g \end{pmatrix}$$

and

$$B(0, 1) + B(1, t) = B(0, t + 1) + B(1, 0).$$

Together with the relation

$$B(1, t) = B(1, 0) + B(0, t) - B(0, 0)$$

we obtain

$$B(0, t+1) = B(0,1) - B(0,0) + B(0,t).$$

By induction we obtain that for each natural number τ

$$B(0,t) = \tau B(0,1) - (\tau - 1)B(0,0),$$

where $t \in k$ denotes the coset of τ in the finite field k. In the case $\tau = p$ we especially have $t = 0$, hence

$$pB(0,0) = pB(0,1).$$

The values of our functions being complex numbers, we obtain

$$B(0,0) = B(0,1).$$

Now we obtain that $B(0,t)$ is constant and

$$B(\alpha, \beta) = B(\alpha, 0).$$

We have to make use of a last admissible quadruple, namely

$$((t+1)f + tg, f + g, (t+1)f + tg + e, f + g + e) \quad (t \in k).$$

We obtain the relation

$$B(t,0) + B(1,0) = B(0,0) + B(t+1,0),$$

which implies in a similar way that $B(t,0)$ and finally that $B(\alpha, \beta)$ is constant.

We generalize the set $\mathcal{P}(1)$ and define for each $t \in k - \{0\}$

$$\mathcal{P}(t) = \left\{ P = \begin{pmatrix} x_1 f + x_2 g + x_3 e \\ y_1 f + y_2 g + y_3 e \end{pmatrix}, \quad t = x_1 y_2 - x_2 y_1 \right\},$$

and

$$\mathcal{P}' = \bigcup_{t \neq 0} \mathcal{P}(t) = \{P; \ x_1 y_2 - x_2 y_1 \neq 0\}.$$

The map

$$\mathcal{P}' \longrightarrow k - \{0\},$$
$$P \longmapsto t(P) := x_1 y_2 - x_2 y_1,$$

is surjective. Applying 4.9 to the basis tf, g, e instead of f, g, e we obtain

4.12 Lemma. *Let*

$$A : \mathcal{P}' \longrightarrow \mathbb{C}$$

be a function with the following two properties:

1) $A \begin{pmatrix} a \\ b \end{pmatrix} = A \begin{pmatrix} a \\ a+b-e \end{pmatrix} = A \begin{pmatrix} a+b-e \\ b \end{pmatrix}$ *for* $\begin{pmatrix} a \\ b \end{pmatrix} \in \mathcal{P}(1).$

2) $A\begin{pmatrix} a \\ b \end{pmatrix} + A\begin{pmatrix} c \\ d \end{pmatrix} = A\begin{pmatrix} a \\ d \end{pmatrix} + A\begin{pmatrix} c \\ b \end{pmatrix}$ *if*

(a)   *a, b, c are linearly independent,*

(b) $d = b + t(a - c)$ *for a $t \in k$,*

(c) $\begin{pmatrix} a \\ b \end{pmatrix}, \begin{pmatrix} c \\ d \end{pmatrix}, \begin{pmatrix} a \\ d \end{pmatrix}, \begin{pmatrix} c \\ d \end{pmatrix}$ *are contained in* $\mathcal{P}$.

Then there exists a function

$$B : k - \{0\} \longrightarrow \mathbb{C},$$

such that

$$A(P) = B(t(P)).$$

We have to investigate the function B in more detail. We can assume $p > 2$, because otherwise $k - \{0\}$ consists of only one element.

4.13 Lemma. *In the case $p > 2$ there exists a unique function*

$$C : k \longrightarrow \mathbb{C},$$

such that

$$B(\alpha) + B(\beta) = C(\alpha + \beta - 2) \quad for \ \alpha, \beta \in k - \{0\}.$$

Proof. Let

$$\alpha, \beta \in k - \{0\} \quad and \quad \gamma \in k.$$

There exist elements $x_1, x_2, x_3 \in k$, such that the vectors

$$a = \alpha f + 0g + x_1 e,$$
$$b = 0f + g + x_2 e,$$
$$c = \beta f + \gamma g + x_3 e$$

are linearly independent. We define

$$d = b + a - c.$$

The quadruple (a, b, c, d) is admissible (i.e. the conditions 2)(a),(b),(c) in 4.9 are satisfied), if the elements

$$\alpha, \beta - \alpha\gamma, \alpha(1 - \gamma), \beta$$

are different from 0 and in this case we have

$$B(\alpha) + B(\beta - \alpha\gamma) = B(\alpha(1 - \gamma)) + B(\beta).$$

This relation implies

$$B(\alpha) + B(\beta) = B(\alpha') + B(\beta') \quad \text{if } \alpha + \beta = \alpha' + \beta' \neq 0$$

and hence 4.13.

We have to determine the solutions of the functional equations 4.13. We may assume

$$B(1) = 0.$$

From 4.12, it follows that
$$B(\alpha) = C(\alpha - 1)$$

and that
$$C(\alpha - 1) + C(\beta - 1) = C(\alpha + \beta - 2) \quad (\alpha, \beta \in k - \{0\})$$

or
$$C(\alpha) + C(\beta) = C(\alpha + \beta) \quad (\alpha, \beta \in k - \{-1\}).$$

By induction we obtain

$$C(t) = \tau \cdot C(1), \quad \tau \in \mathbb{N}, \ 0 \leq \tau < p - 1,$$

where t is the coset of the natural number τ in k. In the special case $\tau = p - 2$, we obtain
$$C(-2) = (p - 2)C(1).$$

Now we assume $p > 3$. Then 2 and -2 are both different from -1 in k and we obtain

$$0 = C(0) = C(2) + C(-2) = 2C(1) + (p - 2)C(1) = pC(1).$$

The characteristic of $\mathbb{C}$ being 0 we obtain $C(1) = 0$ and then $C = 0$. What we have proved is

4.14 Lemma. *If $p \neq 3$, each function*

$$A : \mathcal{P}' \longrightarrow \mathbb{C}$$

with the properties 1) and 2) formulated in 4.12 is constant.

If $p = 3$, the space of solutions A is two-dimensional.

Corollary. *Let $A : \mathcal{P} \to \mathbb{C}$ be a function from the space $\mathcal{W}$, which vanishes on primitive matrices of the form*

$$\begin{pmatrix} a \\ e \end{pmatrix}, \quad \begin{pmatrix} e \\ b \end{pmatrix}.$$

Then the restriction of A to $\mathcal{P}'$ is constant.

Only the corollary in the case $p = 3$ has to be proved. By means of the two quadruples

$$(f - e, f + 2e, g, 2f - g + e), \quad (f - e, f + 2e, -g, 2f + g + e)$$

one obtains the relations

$$A\left(\frac{f - e}{f + 2e}\right) + B(-2) = B(-1) + B(-1),$$

$$A\left(\frac{f - e}{f + 2e}\right) + B(2) = B(1) + B(1).$$

As $-2 = 1$ and $-1 = 2$ in k, $3B(2) = 3B(1)$, and hence $B(1) = B(2)$.

Now we want to enlarge the set $\mathcal{P}'$ by constructing orbits which are not contained completely in $\mathcal{P}'$.

4.15 Definition. *Let $\mathcal{P}''$ be the set consisting of all primitive matrices*

$$P = \begin{pmatrix} a \\ b \end{pmatrix}$$

such that there exist $c \in k^{(1,3)}$ and $t \in k$, such that
  (a) *a, b, c are linearly independent,*
  (b) *$d = b + t(a - c)$,*
  (c) $\begin{pmatrix} c \\ d \end{pmatrix} \begin{pmatrix} a \\ d \end{pmatrix}$ *and* $\begin{pmatrix} c \\ b \end{pmatrix}$ *are contained in $\mathcal{P}'$.*

By definition of the space $\mathcal{W}$, we obtain

4.16 Remark. *In lemma 4.14 the set $\mathcal{P}'$ can be replaced by the bigger set $\mathcal{P}' \cup \mathcal{P}''$.*

The following lemma completes the proof of 4.2.

4.17 Lemma. *Assume $p > 2$. Let $P = \begin{pmatrix} a \\ b \end{pmatrix}$ be a primitive matrix such that neither a nor b is contained in ke. Then*

$$P \in \mathcal{P}''.$$

Proof. Assume $P \notin \mathcal{P}''$ and write

$$a = x_1 f + x_2 g + x_3 e, \quad b = y_1 f + y_2 g + y_3 e,$$

this means

$$(x_1, x_2) \neq (y_1, y_2), \quad x_1 y_2 = x_2 y_1.$$

The group $GL(2, k)$ acts on $\mathcal{P}'$ and $\mathcal{P}''$ by

$$\left(U, \begin{pmatrix} x_1 & x_2 & x_3 \\ y_1 & y_2 & y_3 \end{pmatrix}\right) \longmapsto \begin{pmatrix} x_1 & x_2 & x_3 \\ y_1 & y_2 & y_3 \end{pmatrix} \cdot \begin{pmatrix} U & 0 \\ 0 & 1 \end{pmatrix}.$$

Therefore we may assume

$$a = f + x_3 e, \quad b = y_1 f + y_e \quad (y_1 \neq 0).$$

We define

$$c = g \quad \text{and} \quad d = b + t(a - c)$$
$$= (y_1 + t)f - tg + (y_3 + tx_3)e.$$

The vectors a, b, c are linearly independent. Since $p > 2$ we can choose t such that $y_1 + t$ and t are both different from 0. But then the matrices

$$\begin{pmatrix} c \\ d \end{pmatrix}, \quad \begin{pmatrix} a \\ d \end{pmatrix} \quad \begin{pmatrix} c \\ b \end{pmatrix}$$

are contained in $\mathcal{P}'$.

4.18 Lemma. *Let $A : \mathcal{P} \to \mathbb{C}$ be a function with the properties formulated in 4.6. Then A vanishes on all matrices $\begin{pmatrix} a \\ b \end{pmatrix}$ such that neither a nor b is contained in ke.*

It remains to show

$$A \begin{pmatrix} f \\ f + e \end{pmatrix} = A \begin{pmatrix} f + e \\ f \end{pmatrix} = 0$$

in the case $p = 2$. Both cases can be treated similarly. We consider the first one. By means of the quadruples

$$(f, f + g, g, g + e),$$
$$(f, f + e, f + g, f + g + e),$$
$$(g, g + e, f + g, f + g + e),$$

one obtains

$$A \begin{pmatrix} f \\ f + e \end{pmatrix} + A \begin{pmatrix} g \\ g + e \end{pmatrix} = 0,$$

$$A \begin{pmatrix} f \\ f + e \end{pmatrix} + A \begin{pmatrix} f + g \\ f + g + e \end{pmatrix} = 0,$$

$$A \begin{pmatrix} g \\ g + e \end{pmatrix} + A \begin{pmatrix} f + g \\ f + g + e \end{pmatrix} = 0,$$

and therefore

$$A \begin{pmatrix} f \\ f + e \end{pmatrix} = 0.$$

For the proof of 4.6 it remains to show

$$A \begin{pmatrix} a \\ te \end{pmatrix} = A \begin{pmatrix} te \\ b \end{pmatrix} = 0.$$

By assumption this is true for $a = b = f$. The general case is treated by choosing suitable orbits. By means of the quadruples

$$(te, f, f + te - b, b), \quad t \neq 0$$
$$(b = f - (f + te - b + te))$$

one shows

$$A \begin{pmatrix} te \\ b \end{pmatrix} = 0,$$

if e, f, b are independent. The same argument shows that this is true if e, g, b are independent. But e and b being independent one of the two cases occurs. We obtain

$$A \begin{pmatrix} te \\ b \end{pmatrix} = 0 \quad \text{and similarly} \quad A \begin{pmatrix} a \\ te \end{pmatrix} = 0.$$

Proposition 4.2 is completely proved.

5   An exceptional isotropic structure
The case of a ground field

In this section we investigate a special isotropic structure, which in general does not come from a quadratic form (in the sense 1.10).

Let $R = k$ be a finite field. We identify the vector space k^r with its dual by means of the pairing

$$k^r \times k^r \longrightarrow k,$$
$$(a, b) \longmapsto a'b,$$

similarly for $k^{(r,m)}$. We use the notation

$$e(V, Y) = e(\sigma(V'Y)).$$

5.1 Definition. *A subspace*

$$L \subset k^r$$

is called isotropic if and only if it is different from k^r.

Corollary. *A matrix*

$$V \in k^{(r,m)}, \quad m \geq r,$$

is in this special case isotropic if and only if it is not primitive.

5.2 Proposition. *Assume $n \geq 2r$. The pairs (k^r, n) (equipped with the special isotropic structure 5.1) are admissible.*

We recall the definition of admissibility.

5.3 Definition. *Assume $m \geq r$. A function*

$$f : \mathcal{M} \longrightarrow \mathbb{C}$$

on some subset $\mathcal{M} \subset k^{(r,m)}$ degenerates, if it can be written as linear combination of characters

$$Y \longmapsto e(V, Y), \quad V \ \text{not primitive}.$$

Proposition 5.2 states

5.4 Proposition. *Assume $n \geq 2r$. For a function*

$$f : \mathcal{P}(r, n) \longrightarrow \mathbb{C}$$

on the set of primitive $r \times n$-matrices, the following two conditions are equivalent:
1) f degenerates.
2) The function

$$k^{(r, n-r)} \longrightarrow \mathbb{C},$$
$$Y \longmapsto f((E, Y)U),$$

degenerates for each $U \in \mathrm{GL}(n, k)$.

The direction 1) $\Rightarrow$ 2) is trivial, we have to prove the converse.

The group

$$\mathrm{GL}(r, k) \times \mathrm{GL}(n, k)$$

acts on the set of primitive $r \times n$-matrices by multiplication from both sides. As a consequence this group acts on the set of all functions

$$f : \mathcal{P}(r, n) \longrightarrow \mathbb{C}$$

by

$$f(P) \longmapsto f(A^{-1}PB),$$
$$A \in \mathrm{GL}(r, k), \ B \in \mathrm{GL}(n, k).$$

and also on the subset of functions with the properties 1) resp. 2).

We make use of the action of $\mathrm{GL}(r, k)$. Let

$$\pi : \mathrm{GL}(r, k) \longrightarrow \mathrm{GL}(\mathcal{V})$$

be a representation of this group on a finite dimensional complex vector space $\mathcal{V}$. We investigate vector valued functions

$$f : \mathcal{P}(r, n) \longrightarrow \mathcal{V}$$

with the property

$$f(AP) = \pi(A)f(P) \ \ A \in \mathrm{GL}(r, k).$$

It is sufficient to prove proposition 5.4 for such vector valued functions instead of scalar valued ones. (The condition 5.3 of degeneration is understood to be true for all components of f with respect to a basis of $\mathcal{V}$.) We denote the vector space of all such functions by

$$\mathrm{Ind}(\pi) = \{f : \mathcal{P}(r, n) \longrightarrow \mathcal{V}, \ f(AP) = \pi(A)f(P)\}.$$

The group $\mathrm{GL}(n, k)$ acts on this space by translation from the right. Conditions 1) and 2) in 5.4 are stable under the action of $\mathrm{GL}(n, k)$. Each representation of a finite group on a complex vector space can be decomposed into irreducible constituents. Therefore it is sufficient to prove the following statement.

Let

$$W \subset \text{Ind}(\pi)$$

be a GL(n, k)-submodule of functions with the property 2) in 5.4. If W is not the zero space, it contains a function $f \neq 0$ with the property 1) in 5.4.

Proof. Let $f \in W$ be a non-vanishing function. The function

$$F(P) = F(X,Y) := \sum_{A \in GL(r,k), \quad B \in k^{(r,n-r)}} f((XA, Y + XB)U),$$

$$P = (X,Y); \quad X \in k^{(r,r)}, \quad Y \in k^{(r,n-r)},$$

is contained in the same GL(n, k)-module for each $U \in GL(n, k)$ because of

$$(XA, Y + XB) = (X,Y) \begin{pmatrix} A & B \\ 0 & E \end{pmatrix}.$$

In the special case $X = 0$, we have

$$F(0,Y) = Cf((0,Y)U)$$

with some constant $C \neq 0$. Now we make use of the fact that each primitive matrix P can be written as

$$P = (0,Y)U, \quad Y \in \mathcal{P}(r, n - r).$$

(For this the assumption $n \geq 2r$ is essential!)

For suitable U the function F is different from 0. It has the invariance property

$$F(XA, Y + XB) = F(X,Y).$$

It is therefore sufficient to prove the following lemma.

5.5 Lemma. *Let*

$$f : \mathcal{P}(r, n) \longrightarrow \mathcal{V}$$

be a function with the invariance properties

1) $f(AP) = \pi(A)f(P),$
2) $f(XA, Y + XB) = f(X,Y) \quad (A \in GL(r,k), B \in k^{(r,n-r)}),$
3) f *satisfies condition 2) in 5.4.*

Then f satisfies condition 1) in 5.4.

The proof of 5.4 will be given by induction: We consider for an integer $t \geq 0$ the following

Property $E(t)$.

The function f has property $E(t)$ if and only if

$$f(X, Y) = 0 \quad \text{if} \quad \operatorname{rank} X > t.$$

Beginning of the induction: $t = 0$.

Property $E(0)$ means

$$f(X, Y) = 0 \quad \text{for} \quad X \neq 0.$$

We are going to derive from 5.4, 2) that f vanishes identically. If the matrix $(0, Y)$ is primitive then (X, Y) is primitive for all $X \in k^{(r,r)}$. By assumption the function $G \mapsto f((E, G)U)$, $U \in \operatorname{GL}(n, k)$ degenerates. If $P = (P_1, P_2)$, $P_2 \in k^{(r,r)}$ is an imprimitive matrix then the same is true for P_2. It follows that $X \mapsto f((E, 0, X)U)$, $X \in k^{(r,r)}$ degenerates. Choosing a suitable U, we obtain:

The function

$$k^{(r,r)} \longrightarrow \mathbb{C},$$
$$X \longmapsto f(X, Y)$$

degenerates for each primitive Y. From the well-known orthogonality relations of characters, it follows that

$$f(0, Y) = \sum_X f(X, Y) e(X, E) = 0.$$

Induction step: We consider the case $t > 0$ and assume that 5.4 has been proved for all functions with the property $E(t - 1)$. We are going to prove 5.4 for a function f with the property $E(t)$. It is worthwhile to consider the case $t = r$ seperately. From 5.5, 2) follows that the function $Y \mapsto f(X, Y)$ is constant for invertible X. Hence the induction hypothesis can be applied to the function $g(P) = f(P) - f(E, 0)$. Now we assume

$$0 < t < r.$$

We consider the function

$$h(Q) := f \begin{pmatrix} E^{(t)} & 0 & 0 \\ 0 & 0 & Q \end{pmatrix}, \quad Q \in \mathcal{P}(r - t, n - r).$$

We blow it up to a function

$$H : \mathcal{P}(r - t, n) \longrightarrow \mathcal{V}$$

by

$$H(R, Q) = \begin{cases} 0 & \text{for } R \neq 0, \\ h(Q) & \text{for } R = 0. \end{cases}$$

$(R \in k^{(r-t,r)},\ Q \in k^{(r-t,n-r)})$. We notice that Q is primitive if $(0, Q)$ is primitive. By means of the natural projection

$$P(r, n) \longrightarrow P(r - t, n),$$
$$P = \begin{pmatrix} P_1 \\ P_2 \end{pmatrix} \longmapsto P_2,$$

the function H can be blown up to a function G on the original space $P(r, n)$.

$$G(P) := H(P_2).$$

The function G degenerates as it can be written as a linear combination of characters

$$e(V, P); \quad V = \begin{pmatrix} V_1 \\ V_2 \end{pmatrix}, \quad V_1 = 0.$$

One also has

$$G(X, Y) = 0 \quad \text{if} \ \operatorname{rank} X > t \quad (X \in k^{(r,r)}),$$

because in the decomposition $X = \begin{pmatrix} * \\ R \end{pmatrix}$ the matrix R will be different from 0.

To enforce the invariance properties formulated in 5.5 we symmetrize G:

$$F(P) = F(X, Y) := \sum_{A, \tilde{A} \in \mathrm{GL}(r,k);\ B \in k^{(r,n-r)}} \pi(A)^{-1} G(A(X\tilde{A}, Y + XB)).$$

5.6 Lemma. *The function F has the following properties:*
1) F degenerates.
2) F has the invariance properties formulated in 5.4.
3) $F(X, Y) = 0$ if rank $X > t$.

Proof. 1) is a consequence of the same property of G, because F is contained in the $\mathrm{GL}(r, k) \times \mathrm{GL}(n, k)$-module generated by G.

2) The invariance properties are true by the construction of F.

3) The ranks of $AX\tilde{A}$ and X are the same. Hence F inherits the claimed property from G.

It is sufficient to prove 5.4 for the function

$$\tilde{f} = f + CF$$

instead of f, where C denotes an arbitrary number. This follows from the induction hypothesis and

5.7 Lemma. *There exists a complex number C such that the function $\tilde{f} = f + CF$ satisfies proerty $E(t - 1)$.*

Because f and F both satisfy $E(t)$, this means

$$\tilde{f}(X, Y) = 0 \quad \text{if} \quad \text{rank } X = t.$$

If rank $X = t$, we find matrices $A, \tilde{A} \in \text{GL}(r, k)$, such that

$$AX\tilde{A} = \begin{pmatrix} E^{(t)} & 0 \\ 0 & 0 \end{pmatrix}.$$

After that we obtain by suitable choice of B

$$A(X\tilde{A}, Y + XB) = \begin{pmatrix} E^{(t)} & 0 & 0 \\ 0 & 0 & Q \end{pmatrix}.$$

the proof of 5.4 will be complete after the proof of

5.8 Lemma. *The functions f and F coincide on the set of matrices*

$$P = \begin{pmatrix} E^{(t)} & 0 & 0 \\ 0 & 0 & Q \end{pmatrix}, \quad Q \in \mathcal{P}(r - t, n - r),$$

up to a constant factor.

Proof. We have to decompose a matrix

$$X \in k^{(r,m)} \quad (m \quad \text{variable})$$

several times into blocs

$$X = \begin{pmatrix} X_1 \\ X_2 \end{pmatrix}; \quad X_2 \in k^{(r-t,m)}$$

and call X_2 "the second row" of X. Our goal is to compute the function $F(X, Y)$ for

$$X = \begin{pmatrix} E^{(t)} & 0 \\ 0 & 0 \end{pmatrix}.$$

In the definition of $F(X, Y)$ only values $G(AX\tilde{A}, *)$ occur. By definition of G they vanish unless the second row of $AX\tilde{A}$ vanishes. An easy calculation shows:
If the second row of $A \begin{pmatrix} E^{(t)} & 0 \\ 0 & 0 \end{pmatrix} \tilde{A}$ ($A, \tilde{A} \in \text{GL}(r, k)$) vanishes, then A is of the form

$$A = \begin{pmatrix} a & b \\ 0 & d \end{pmatrix}, \quad d \in \text{GL}(r - t, k)$$

and in this case the second row of the matrix

$$AY + A \begin{pmatrix} E^{(t)} & 0 \\ 0 & 0 \end{pmatrix} B$$

equals the second row of AY. If we denote by Q the second row of Y, the second row of AY equals

$$dQ \quad (Q \text{ the second row of } Y).$$

We have shown that $F \begin{pmatrix} E & 0 & 0 \\ 0 & 0 & Q \end{pmatrix}$ is a sum of terms of the form

$$\pi \begin{pmatrix} a & b \\ 0 & d \end{pmatrix}^{-1} h(dQ) = \pi \begin{pmatrix} a & b \\ 0 & d \end{pmatrix}^{-1} f \begin{pmatrix} E & 0 & 0 \\ 0 & 0 & dQ \end{pmatrix}.$$

Using the transformation property of f under $\mathrm{GL}(r, k)$ (translation from the left) we see that all those terms are of the form

$$f \begin{pmatrix} a^{-1} & 0 & * \\ 0 & 0 & Q \end{pmatrix}.$$

Using the invariance of f under the action $Y \to Y + XB$, we obtain that the last term equals

$$f \begin{pmatrix} a^{-1} & 0 & 0 \\ 0 & 0 & Q \end{pmatrix}.$$

6   An exceptional isotropic structure
The case of a ground ring

In this section we generalize the results of sec.5 to the case of a local ground ring, more precisely: We will reduce the general case to the case of a ground field.

Let R be a finite local principal ideal ring with maximal ideal $\mathbf{m}$ and residue field

$$k = R/\mathbf{m}.$$

We have to consider a finitely generated module M together with its dual module M^*. In this section we generalize the exceptional isotropic structure of the last section:

6.1 Definition. *A submodule*

$$L \subset M^*$$

is called isotropic, if and only if it is properly contained in M^*.

By the Nakayama lemma for a submodule $L \subset M^*$ one has

$$L = M^* \iff L \otimes_R k = M^* \otimes_R k.$$

6.2 Remark. *A vector*

$$V \in (M^*)^m$$

is isotropic if and only it is imprimitive mod $\mathbf{m}$.

In this section we will prove:

6.3 Proposition. *Let r be the rank of M (i.e. the dimension of $M \otimes k$). Assume $n \geq 2r$. The pair (M, n) (equipped with the special isotropic structure 6.1) is admissible.*

We recall the definition of admissibility.

6.4 Definition. *Assume $m \geq r$. A function*

$$f : \mathcal{M} \longrightarrow \mathbb{C}$$

on some subset $\mathcal{M} \subset M^m$ degenerates, if it can be written as linear combination of characters

$$Y \longmapsto e(V, Y), \quad V \text{ not primitive.}$$

We recall that we have chosen a non-trivial additive character

$$e : R \longrightarrow \mathbb{C}$$

and that we defined

$$e(V, Y) = e(V(Y)) \quad \text{for } Y \in M^m, \ V \in (M^*)^m \ (= (M^m)^*).$$

Proposition 6.3 states

6.5 Proposition. *Assume $n \geq 2r$. For a function*

$$f : \mathcal{P}(M, n) \longrightarrow \mathbb{C}$$

on the set of primitive vectors in M^n, the following two conditions are equivalent.
1) f degenerates.
2) For

$$E = (e_1, \ldots, e_r),$$

a minimal system of generators of M, the function

$$M^{n-r} \longrightarrow \mathbb{C},$$
$$Y \longmapsto f((E, Y)U)$$

degenerates for each $U \in \mathrm{GL}(n, k)$.

As a first step, we prove the following

6.6 Lemma. *Let*

$$f : M^m \longrightarrow \mathbb{C}; \quad m \geq r \quad (r = \mathrm{rank}\ M),$$

be a function, such that the function

$$f_{U,A} : M^r \longrightarrow \mathbb{C}$$
$$f_{U,A}(Y) = f((A, Y)U)$$

degenerates for all $A \in M^{m-r}$ and all $U \in \mathrm{GL}(m, R)$. Then f degenerates too.

Proof. We compare the (unique!) expansions as linear combination of characters

$$f(X) = \sum_{V \in (M^*)^m} C(V)e(V, X),$$

$$f_{U,A}(Y) = \sum_{W \in (M^*)^r} C_{U,A}(W)e(W, Y).$$

One has

$$C_{U,A}(W) = \sum_{G} C((G, W)U'^{-1})e(G, A).$$

By assumption $C_{U,A}(W)$ vanishes for primitive W. Because A is variable, $C((*, W)U)$ has to vanish for primitive W and $U \in GL(m, R)$. Each primitive $P \in M^n$ can be written in the form

$$P = (G, W)U, \qquad W \text{ primitive}.$$

This gives us $C(P) = 0$ for primitive P.

An immediate consequence of 6.6 is

6.7 Lemma. *Let*

$$E = (e_1, \ldots e_r)$$

be a minimal system of generators of the $R-$module M. For a function

$$f : \mathcal{P}(M, n) \longrightarrow \mathbb{C} \qquad (n \geq 2r)$$

the following statements are equivalent:

1) The function

$$M^{n-r} \longrightarrow \mathbb{C},$$

$$X \longmapsto f((E, X)U),$$

   degenerates for each $U \in GL(n, R)$.

2) The function

$$M^r \longrightarrow \mathbb{C}$$

$$Y \longmapsto f((E, 0, Y)U),$$

   degenerates for all $U \in GL(n, R)$.

   (0 denotes the zero element in M^{n-2r}.)

Now we want to reduce proposition 6.5 to the case of a ground field.

We denote by p a generator of the maximal ideal $\mathbf{m}$ of R. Let $P \in M^m$ be a primitive $m-$tuple. Then

$$P + pX, \quad X \in M^m,$$

is primitive too. The space of all functions

$$f : \mathcal{P}(M, n) \longrightarrow \mathbb{C}$$

can be decomposed into eigen spaces of the translation module pM^n. Properties 1) and 2) of proposition 6.5 being invariant under translation, it is sufficient to restrict to eigen functions.

Let f be a function with the following two properties:

a) $f(P+X) = \chi(X)f(P)$ for $X \in pM^n$.

b) f satisfies condition 2) of proposition 6.5.

Here

$$\chi : pM^n \longrightarrow \mathbb{C}^\bullet$$

denotes some character.

We have to show that f satisfies the (stronger) condition 1) of proposition 6.5.

Because of the exactness of the dualizing functor, the character χ can be written as the restriction of a character on M^n,

$$\widetilde{\chi} : M^n \longrightarrow \mathbb{C}^\bullet.$$

Instead of f we consider the function

$$\varphi(P) = \widetilde{\chi}(-P)f(P).$$

Obviously

$$\varphi(P + X) = \varphi(P) \quad \text{for } X \in pM^n.$$

Notation:

$$\overline{M} = M/\mathbf{m}M = M/pM.$$

$\overline{M}$ is a vector space over the field k. The canonical map

$$\mathcal{P}(M, n) \longrightarrow \mathcal{P}(\overline{M}, n),$$
$$P \longmapsto \overline{P}.$$

is surjective. The function φ is induced by a function

$$\Phi : \mathcal{P}(\overline{M}, n) \longrightarrow \mathbb{C},$$
$$\Phi(\overline{P}) = \varphi(P).$$

We want to rewrite condition 2) for the function Φ. For this purpose we write the character $\widetilde{\chi}$ in the form

$$\widetilde{\chi}(X) = e(A, X), \quad A \in (M^*)^n.$$

Condition 2) states:

$$\varphi((E, Y)U) = e(-A, (E, Y)U) \cdot \sum_{V \text{ not primitive}} C_U(V)e(V, Y).$$

We decompose AU' into two parts

$$AU' = (*, (AU')_2), \quad (AU')_2 \in (M^*)^{n-r},$$

and obtain

$$e(-A, (E, Y)U) = e(-AU', (E, Y)) = C \cdot e(-(AU')_2, Y),$$

where C is a constant independent of Y. This gives us

$$\varphi((E, Y)U) = C \sum_{W + (AU')_2 \text{ not primitive}} C'_U(W) e(W, Y),$$

where $C'_U(W) = C_U(W + (AU')_2)$. Because of the periodicity of the function φ we may replace $e(W, Y)$ in this formula by

$$e_0(W, Y) = \sharp(pM^{n-r})^{-1} \sum_{X \in pM^{n-r}} e(W, Y + X).$$

This expression is different from 0 only if W is contained in the orthogonal complement of pM^{n-r}:

$$W \in (pM^{n-r})^{\perp} = \{L \in (M^*)^{n-r}; \quad L | pM^{n-r} = 0\}.$$

Let N be an arbitrary finitely generated module (for example $N = M^{n-r}$). The orthogonal complement of pN can be identified with the dual of the vector space $\overline{N} = N \otimes k$. More precisely: The canonical projection $N \to \overline{N}$ induces by dualizing an imbedding

$$\operatorname{Hom}_k(\overline{N}, k) = \operatorname{Hom}_R(\overline{N}, R) \hookrightarrow \operatorname{Hom}_R(N, R),$$

whose image is the orthogonal complement of pN. We obtain an isomorphism

$$(pN)^{\perp} \xrightarrow{\sim} (\overline{N})^* = \operatorname{Hom}_k(\overline{N}, k)$$
$$W \longmapsto \overline{W}.$$

Now we can interpret e_0 as a pairing

$$e_k : (\overline{M}^*)^{n-r} \times \overline{M}^{n-r} \longrightarrow \mathbb{C},$$
$$e_k(\overline{W}, \overline{Y}) := e(W, Y).$$

From the definition of $e(W, Y)$ it is clear that this pairing depends only on $\overline{W}(Y)$. Therefore it is given by a function

$$e_k : k \longrightarrow \mathbb{C}^{\bullet}.$$

Obviously e_k is a non-trivial additive character.

Now we can rewrite conditions 1) and 2) from 6.5 for the function Φ:

6.8 Lemma. *Let $U \in \mathrm{GL}(n, R)$. We write the function*

$$k^{(r, n-r)} \longrightarrow \mathbb{C},$$
$$Y \longmapsto \Phi((E, Y)U),$$

as a linear combination of characters,

$$\Phi((E, Y)U) = \sum_{V \in (\overline{M}^*)^{n-r}} D_U(V) e_k(V, Y).$$

The condition 2) in proposition 6.5 for the function Φ implies that the coefficient $D_U(V)$ can be different from 0 only if there exists a representative $W \in (M^*)^{n-r}$ of V ($V = \widetilde{W}$), such that

$$W + (AU)_2 \text{ is not primitive and such that } \quad W|pM^{n-r} = 0.$$

The expressions $\Phi((E, Y)U)$, $C_U(V)$ and the condition formulated in 6.8 only depend on the image of U in $\mathrm{GL}(n, k)$. But to formulate the condition, we need a representative in $\mathrm{GL}(n, R)$. This group acts on each k-module by means of the natural projection to $\mathrm{GL}(n, k)$.

It is worth while to reformulate also condition 1) in 6.5 for the function Φ.

6.9 Lemma. *The condition 1) in proposition 6.5 for the function Φ implies that the function Φ is a linear combination of functions*

$$P \longmapsto e(V, P),$$
$$V = \overline{W}; \quad W + A \text{ not primitive } .$$
$$(W \in (M^*)^n, \quad W|pM^{n-r} = 0).$$

We have to prove that the property formulated in 6.8 implies the one formulated in 6.9.

We make use of the theorem of elementary divisors.

The module M is direct product of monic modules:

$$M \cong \prod_{\nu=1}^{h} R/p^{n_\nu} R.$$

We assume that the numbers $n_\nu > 0$ have been chosen minimal. Then they are unique up to their order. It is useful to consider two special cases seperately:

1. case:

$$n_\nu = 1 \quad \text{for} \quad 1 \le \nu \le h.$$

In this case $M = \overline{M}$ a priori is a vector space over k. In that case the choice $A = 0$ is possible. The statement reduces to the case of a ground field $R = k$. This case has been treated in the previous section.

2. case:

$$n_\nu > 1 \quad \text{for} \quad 1 \le \nu \le h.$$

Claim. *Assume $W \in (M^*)^m$. Then*

$$W|pM^m = 0 \Longrightarrow W \in p(M^*)^m.$$

proof of the claim: We may assume that M is monic and that $m = 1$:

$$M = R/p^t R, \quad t > 1.$$

Each linear form W on M is given by multiplication with an element a of R, $W(m) = am$. The assumption implies

$$pa \equiv 0 \bmod p^t,$$

hence $a \equiv 0 \bmod p^{t-1}$.

The condition $W \in p(M^*)^m$ implies for each $B \in (M^*)^m$:

$$B + W \text{ primitive} \Longleftrightarrow B \text{ primitive}.$$

Hence the condition formulated in 6.9 means nothing else but

$$A \text{ primitive} \Longrightarrow \Phi = 0.$$

Notice: The condition $V|M^n$ means nothing for the coset $\overline{M} \bmod p$.)

This has to be proved by means of the condition formulated in 6.8 which now means:

$$(AU)_2 \text{ primitive} \Longrightarrow \Phi((E,Y)U) = 0.$$

For the proof we can assume that A is primitive, without loss of generality $A = (E, 0)$. The proof of the second case will be complete with the following

6.10 Lemma. *Each primitive matrix*

$$P \in k^{(r,n)}$$

can be written in the form

$$P = (E,Y)U; \quad U = \begin{pmatrix} U_1 & U_2 \\ U_3 & U_4 \end{pmatrix},$$

where the subbloc U_2 is primitive.

Proof. The group $GL(n,k)$ acts transitively on the space of primitive matrices. Hence we can write

$$P = (E,0)V, \quad V \in GL(n,k).$$

It follows

$$P = (E,Y) \begin{pmatrix} E & -Y \\ 0 & E \end{pmatrix} V.$$

We have to choose Y such that the subbloc U_2 of the matrix

$$U = \begin{pmatrix} E & -Y \\ 0 & E \end{pmatrix} V$$

is primitive. This subbloc equals

$$U_2 = V_2 - YU_4, \quad V = \begin{pmatrix} V_1 & V_2 \\ V_3 & V_4 \end{pmatrix}.$$

It is easy to find a matrix Y with the desired property.

We still have to treat the mixed case. By the theorem of elementary divisors it is possible to decompose the module M as follows:

$$M = M_1 \times M_2,$$

where

a) M_1 is a k-module $(pM_1 = 0)$.

b) No composition factor of M_2 is a $k-$module.

We can assume that M_1 and M_2 are both different from 0, because we have treated already the pure cases. We decompose vectors $P \in M^n$ into pairs

$$P = (P^{(1)}, P^{(2)}); \quad P^{(1)} \in M_1^n; \; P^{(2)} \in M_2^n.$$

The matrix P is primitive, if and only if both components $P^{(1)}, P^{(2)}$ are primitive. In the decomposition of the matrix $A = (A^{(1)}, A^{(2)})$ we may assume $A^{(1)} = 0$.

3. case (mixed case) a) $A^{(2)}$ is primitive

In this case a matrix $W + A$ under the assumption $W|pM^n = 0$ is not primitive if and only if $W^{(1)}$ is not primitive. There is no condition for $W^{(2)}$. Hence the statement is:

For each fixed $P^{(2)}$ the function

$$P^{(1)} \longmapsto \Phi(P^{(1)}, P^{(2)})$$

degenerates.

To prove this, we write the fixed matrix $P^{(2)}$ in the form

$$P^{(2)} = (E^{(2)}, Y^{(2)})U^{(2)}, \quad (A^{(2)}U^{(2)})_2 \; \text{primitive}.$$

This is possible by the discussion of the second case. Then we consider $\Phi((E,Y)U)$, $Y = (Y^{(1)}, Y^{(2)})$, $U = (U^{(1)}, U^{(2)})$. Here $Y^{(2)}$ and $U^{(2)}$ are fixed. From the assumption on Φ it follows that

$$Y^{(1)} \longmapsto \Phi((E,Y^{(1)})U^{(1)}, (E,Y^{(2)})U^{(2)})$$

degenerates for each $U^{(2)}$. The assertion follows from the first case.

3. case (mixed case) b) $A^{(2)}$ is not primitive

In this case there is nothing to be proved as

$$W + A \quad \text{not primitive for } W|pM^n$$

is empty, and hence there is no restriction on the function Φ.

All the statements of this section have now been proved.

7   The case $n \geq 2r$ in general

We are going to show that the case $n \geq 2r$ over an arbitrary ring can be reduced to the exceptional isotropic structure.

7.1 Proposition. *Let M be a pure module of rank r, which has been equipped with an isotropic structure. In the case*

$$n \geq 2r$$

each pair (M, n) is admissable.

Proof. We may assume that the ground ring R is a local ring with maximal ideal $\mathbf{m}$.

We consider a function

$$f : \mathcal{P}(M, n) \longrightarrow \mathbb{C},$$

such that for each $U \in GL(n, R)$ the function

$$f_U : R^{(r, n-r)} \longrightarrow \mathbb{C}$$
$$f_U(Y) = f((E,Y)U')^{-1}$$

is a linear combination of the characters

$$Y \longmapsto e(V, Y), \quad V \text{ isotropic.}$$

We write f in the form

$$f(P) = \sum_V C(V) e(V, P), \quad C(V) \in \mathbb{C}.$$

Such an expression is not unique!

We want to prove 7.1 indirectly and assume:

For each representation of f as a linear combination of characters there exists an anisotropic V, such that $C(V) \neq 0$.

For each vector

$$V = (v_1, \ldots, v_n) \in (M^*)^n$$

we denote by $\langle V \rangle$ the submodule of M, which is generated by $v_1, \ldots v_n$, and by

$$\operatorname{ord} V := \operatorname{rank}\langle V \rangle$$

its rank. The maximal occuring order of an anistropic V in the given representation of f is denoted by

$$r' = \max\{\operatorname{ord} V; \quad V \text{ anisotropic}, \ C(V) \neq 0\}.$$

Under all representations of f as linear combination of characters we choose one such that r' is minimal and after that one such that the number of elements of the set

$$\mathcal{A} := \{V \in (M^*)^n \ \text{anisotropic}, \quad C(V) \neq 0, \quad \operatorname{ord} V = r'\}$$

is minimal.

We fix a vector

$$V_0 \in \mathcal{A}$$

and consider the generated sub-module $\langle V_0 \rangle \subset M^*$. Because of the exactness of the dualizing functor, there is a one-to-one correspondence between submodules of M^* and the factor modules of M. We denote the factor module of M, which corresponds to $\langle V_0 \rangle$, by N;

$$M \to N \to 0 \quad \text{(exact sequence)},$$
$$N^* = \langle V_0 \rangle \subset M^*.$$

We consider the subset

$$\mathcal{B} := \{V \in \mathcal{A}; \quad \langle V \rangle \subset N^*\}.$$

We have to consider the following partial sum of f:

$$g(P) = \sum_{V \in \mathcal{B}} C(V) e(V, P)$$

This function is defined on N^n; more precisely, there exists a function

$$\tilde{g} : N^n \longrightarrow \mathbb{C},$$

such that the diagram

$$\begin{array}{ccc} M^n & \to & \mathbb{C} \\ \downarrow & & \downarrow \\ N^n & \to & \mathbb{C} \end{array}$$

commutes.

Our aim is to replace the given representation of f as linear combination of characters by a new one, in which either the number of elements of $\mathcal{A}$ or the number r' decreases. This will contradict the choice of r' and $\mathcal{A}$. For the construction of the new representation we will have to apply the exceptional case in sec.6 to $\tilde{g}$.

We denote by $\bar{a}$ the image of an elment $a \in M$ in the quotient N. We choose a minimal system of generators

$$E = (e_1, \ldots, e_r)$$

of M, such that

$$\overline{E} = (\bar{e}_1, \ldots, \bar{e}_{r'}) \quad (r' = \operatorname{rank} N)$$

is a minimal system of generators of N and such that

$$\bar{e}_{r'+1} = \ldots = \bar{e}_r = 0.$$

7.2 Lemma. *In the representation*

$$g((E, Y)U) = \sum_{W \in M^{n-r}} C'_U(W) e(W, Y)$$

(which is uniquely determined!), we have

$$C'_U(W) \neq 0 \Longrightarrow \langle W \rangle \neq N^*.$$

Proof. We may assume $U = E$. By assumption, we have

$$f(P) = \sum C(V) e(V, P),$$

$$f((E, Y)) = \sum_{W \text{ isotropic}} C_E(W) e(W, Y)$$

and

$$g(P) = \sum_{V \in \mathcal{B}} C(V) e(V, P).$$

This gives us

$$C_E(W) = \sum_{V=(V_1, W)} C(V) e(V_1),$$

$$C'_E(W) = \sum_{V=(V_1, W) \in \mathcal{B}} C(V) e(V_1).$$

or

$$C'_E(W) = C_E(W) - \sum_{V=(V_1, W) \notin \mathcal{B}} C(V) e(V_1).$$

We have to analyze the condition $C'_E(W) \neq 0$ and distinguish between two cases:

case 1. W is isotropic.

The claim follows from the fact that N^* is anisotropic.

case 2. W is anisotropic.

Then $C_E(W) = 0$ and there exists a matrix V_1 such that

$$V = (V_1, W) \notin \mathcal{B}, \quad C(V) \neq 0.$$

We argue indirectly and assume

$$\operatorname{ord} W = r'.$$

The matrix V is anistropic and $C(V)$ is different from 0. Hence $\operatorname{ord} V \leq r'$ and –by the assumption on W– we have

$$\operatorname{ord} V = \operatorname{ord} W, \quad \text{hence} \quad \langle V \rangle = \langle W \rangle.$$

But V is not contained in $\mathcal{B}$. Therefore $\langle V \rangle$ (and $\langle W \rangle$) are different from N^*.

This completes the proof of 7.2. We want to reformulate 7.2 for $\widetilde{g}$ instead of g. For this purpose we choose a minimal system of generators

$$E = (e_1, \ldots, e_r)$$

with the following two properties:

a) $\overline{E} = (\overline{e}_1, \ldots, \overline{e}_{r'})$ is a minimal system of generators of N.

b) $\overline{e}_{r'+1} = \ldots = \overline{e}_r = 0$. Let

$$F = (e_1, \ldots e_{r'}).$$

From the construction of $\widetilde{g}$ we know

$$\widetilde{g}((\overline{F}, 0, \overline{Y})U) = g((E, Y)U).$$

Lemma 7.2 implies

$$\widetilde{g}((\overline{F}, 0, Y)U) = \sum_{W \in (N*)^{n-r}} \widetilde{C}'_U(W) e(W, Y)$$

$$\widetilde{C}'_U(W) \neq 0 \Longrightarrow \langle W \rangle \neq N^*.$$

From lemma 6.6 and the fundamental lemma in the exceptional case (proposition 6.5) we can conclude

7.3 Lemma. *The function $\widetilde{g}$ is a linear combination of characters*

$$P \mapsto e(V, P); \quad V \in (N^*)^n, \quad \langle V \rangle \neq N^*.$$

Now we reformulate this result again for g and obtain a representation

$$g(P) = \sum_{\operatorname{ord} V < r'} C'(V) e(V, P).$$

The function g was defined as a partial sum of f. Replacing this partial sum by the above expression of g we obtain a new representation of f as linear combination of characters. In this new representation either the number of elements of $\mathcal{A}$ or the number r' decreased. But this is a contradiction to the minimality assumptions on $\mathcal{A}$ and r'.

VI   The results

1   The representation theorem and theta relations

We collect the main results about representation of singular modular forms as linear combination of theta series and point out the connection with the theory of theta relations.

We briefly recall the definition of the "big singular space" (s.IV sec.1).

We fix three natural numbers

$$n \quad (= \text{degree}),$$
$$r \quad (= 2 \cdot \text{weight}),$$
$$q \quad (= \text{level}),$$

and assume

$$r < n \quad (\text{singular case}).$$

Let furthermore

$$\varrho_0 : \text{GL}(n, \mathbb{C}) \longrightarrow \text{GL}(\mathcal{Z}), \quad \dim_{\mathbb{C}} \mathcal{Z} < \infty,$$

be an irreducible reduced representation on a finite dimensional complex vector space. ("Reduced" means that ϱ_0 is polynomial but does not vanish identically on the determinant surface "$\det A = 0$".)

We denote by

$$\mathbf{P} = \mathbf{P}(n, r, q, \varrho_0)$$

the space of all Fourier series

$$f(Z) = \sum_{T=T' \text{ integral}} a(T) \exp \frac{\pi i}{q} \sigma(TZ), \quad a(T) \in \mathcal{Z},$$

converging on $\mathbb{H}_n$ and with the following properties:

a) $a(T) \neq 0 \Longrightarrow T \geq 0$ (positive semidefinite),
b) $a(T) \neq 0 \Longrightarrow \text{rank}\,(T) \leq r$,
c) $a(T[U]) = \varrho_0(U')a(T)$ for all $U \in \text{SL}(n, \mathbb{Z}), U \equiv E \mod q$

The space $\mathbf{P}$ is filtered with respect to r. We are more interested in the successive quotients than in the space $\mathbf{P}$ itself.

$$\overline{\mathbf{P}} = \overline{\mathbf{P}(n, r, q, \varrho_0)} = \frac{\mathbf{P}(n, r, q, \varrho_0)}{\mathbf{P}(n, r-1, q, \varrho_0)}$$

The so called "big singular space" is a certain subspace of $\overline{\mathbf{P}}$. For its definition one has to introduce the group

$$\Gamma = \Gamma_{n,\vartheta}(r),$$

which is the subgroup of the Siegel modular group $\Gamma_n = \mathrm{Sp}(n, \mathbb{Z})$ generated by

a) $$\begin{pmatrix} U' & 0 \\ 0 & U^{-1} \end{pmatrix}, \quad U \in Sl(n, \mathbb{Z}),$$

b)   the so-called "imbedded involution"

$$I_r = \begin{pmatrix} E_r & E - E_r \\ E_r - E & E_r \end{pmatrix}, \quad E_r = E_r^{(n)} = \begin{pmatrix} E^{(r)} & 0 \\ 0 & 0 \end{pmatrix},$$

i.e.

$$I_r = \begin{pmatrix} E^{(r)} & 0 & 0 & 0 \\ 0 & 0 & 0 & E^{(n-r)} \\ 0 & 0 & E^{(r)} & 0 \\ 0 & -E^{(n-r)} & 0 & 0 \end{pmatrix}.$$

The space

$$\mathbf{M} = \mathbf{M}(n, r, q, \varrho_0)$$

consists of all holomorphic functions

$$f : \mathbb{H}_n \longrightarrow \mathcal{Z},$$

such that the function

$$Z \longmapsto \det (CZ + D)^{-r/2} \varrho_0 (CZ + D)^{-1} f(M\, Z)$$

is contained in the space $\mathbf{P}$ for all

$$M = \begin{pmatrix} A & B \\ C & D \end{pmatrix} \in \Gamma_{n,\vartheta}(r).$$

The big singular space is the image

$$\overline{\mathbf{M}} = \overline{\mathbf{M}(n, r, q, \varrho_0)}$$

of $\mathbf{M}$ in $\overline{\mathbf{P}}$.

We consider theta series of the type

$$\sum_{G=G^{(r,n)} \text{ integral}} P(S^{1/2}G) \exp \frac{\pi i}{q} \sigma\{S[G]Z + 2V'G\}.$$

Here $S = S^{(r)}$ is a positive definite matrix, V is a matrix of type $V = V^{(r,n)}$ and P is a polynomial,

$$P : \mathbb{C}^{(r,n)} \longrightarrow \mathcal{Z}.$$

We make the following assumptions on S, V, P.

1)   The matrix $S = S^{(r)}$ is integral and $q^2 S^{-1}$ is half integral if $n = r + 1$ and integral if $n > r + 1$.

An integral matrix $H = H^{(r,n-r)}$ is called **isotropic** with respect to S, if

$$S^{-1}[H + qX]$$

is integral for all integral $X = X^{(r,n-r)}$.

An integral matrix $V = V^{(r,n)}$ is called isotropic, if for every matrix $U \in \mathrm{GL}(n, \mathbb{Z})$ the matrix H defined by

$$VU = (*, H), \qquad H = H^{(r,n-r)}$$

is isotropic.

We assume furthermore

2) $V = V^{(r,n)}$ is isotropic.

3) P is a harmonic form with respect to ϱ_0.

Then the above theta series is contained in the space **M**.

Problem. *Is* $\overline{\mathbf{M}}$ *generated by the (images of the) above theta series?*

In chapter IV we formulated an elementary statement, called the fundamental lemma, which –if it is true– gives an affirmative answer to this problem. In chapter V we proved it in many cases.

The lemma also allows to describe **all linear relations between the involved theta series**. Hence in principle we obtain dimension formulae for **M**.

We are going to describe the results in detail and fix for a moment one of the positive matrices S.

1.1 Definition. *A function*

$$\varphi : \mathbb{Z}^{(r,n)} \longrightarrow \mathbb{C}$$

is called **admissible***, if it is a linear combination of functions of the form*

$$X \longmapsto \exp\left(\frac{2\pi i}{q} \sigma(X'V)\right)$$

with isotropic V.

We can consider φ as a function

$$\varphi : (\mathbb{Z}/q\mathbb{Z})^{(r,n)} \longrightarrow \mathbb{C}.$$

1.2 Remark. *Let* S *be a positive definite* $r \times r-$*matrix such that*

$$S \text{ and } q^2 S^{-1}$$

both are integral. Then the theta series

$$\vartheta_{P,\varphi}(S; Z) = \sum P(S^{1/2}G)\varphi(G) \exp\left(\frac{\pi i}{q}\sigma(S[G]Z)\right)$$

is contained in **M** *for each admissible function* φ *and for each harmonic form* P *with respect to* ϱ_0.

We use the notation

$$\mathbf{H}(\varrho_0)$$

for the space of all harmonic forms $P : \mathbb{C}^{(r,n)} \to \mathcal{Z}$ with respect to ϱ_0. Let

$$\mathbf{A}(S) \subset \{\varphi : (\mathbb{Z}/q\mathbb{Z})^{(r,n)} \longrightarrow \mathbb{C}\}$$

be the subset of all **admissible** functions with respect to (S, q) (Definition 1.1).

We have to consider the restriction of admissible functions to the set of **primitive** matrices

$$\mathcal{P} = \mathcal{P}(r, n) = \{P \in (\mathbb{Z}/q\mathbb{Z})^{(r,n)}; \quad P \text{ primitive}\}.$$

We denote the space of restricted admissible functions by

$$\mathbf{B}(S) = \{\varphi | \mathcal{P}; \quad \varphi \in \mathbf{A}(S)\}.$$

The unit group of S

$$\mathcal{E}(S) := \{U \in \mathrm{GL}(r, \mathbb{Z}); \quad S[U] = S\}$$

acts on harmonic polynomials and admissible functions:

$$P(S^{1/2}G) \longmapsto P(S^{1/2}UG), \qquad \varphi(G) \longmapsto \varphi(UG).$$

We are interested in the invariant subspaces

$$\mathbf{H}(\varrho_0)^{\mathcal{E}(S)}, \quad \mathbf{A}(S)^{\mathcal{E}(S)}, \qquad \mathbf{B}(S)^{\mathcal{E}(S)}.$$

Theta relations arise from the fact that the restriction map

$$\mathbf{A}(S)^{\mathcal{E}(S)} \longrightarrow \mathbf{B}(S)^{\mathcal{E}(S)}$$

is not injective in general.

We choose a subspace

$$\widetilde{\mathbf{A}}(S) \subset \mathbf{A}(S)^{\mathcal{E}(S)},$$

which maps isomorphically to $\mathbf{B}(S)^{\mathcal{E}(S)}$.

$$\widetilde{\mathbf{A}}(S) \xrightarrow{\sim} \mathbf{B}(S)^{\mathcal{E}(S)}.$$

After that we consider the spaces of theta series

$$\Theta(S, q) = \{\vartheta_{P,\varphi}(S; Z); \quad \varphi \in \mathbf{A}(S), \ P \in \mathbf{H}(\varrho_0)\}$$

and

$$\widetilde{\Theta}(S, q) = \{\vartheta_{P,\varphi}(S; Z); \quad \varphi \in \widetilde{\mathbf{A}}(S), \ P \in \mathbf{H}(\varrho_0)^{\mathcal{E}(S)}\}.$$

In the subspace $\widetilde{\Theta}(S, q)$ no theta relations occur:

1.3 Remark. *The map*

$$\widetilde{A}(S) \otimes H(\varrho_0)^{\mathcal{E}(S)} \longrightarrow \widetilde{\Theta}(S, q),$$
$$(\varphi, P) \longmapsto \vartheta_{S,P}(Z; \varphi)$$

is an isomorphism. Especially

$$\dim \widetilde{\Theta}(S, q) = \dim B(S)^{\mathcal{E}(S)} \cdot \dim H(\varrho_0)^{\mathcal{E}(S)}.$$

Proof. It is sufficient to show that the map is injective. This is easily done by considering the Fourier coefficients of the theta series at matrices of the type

$$S[P]; \quad P \text{ primitive.}$$

In general the space $\widetilde{\Theta}(S, q)$ is a proper subspace of $\Theta(S, q)$. But nevertheless one can prove the following refined version of a representation theorem.

1.4 Theorem. *Let S run through a set of representatives of unimodular classes of all positive integral matrices S such that $q^2 S^{-1}$ is half integral if $n = r + 1$ and integral if $n > r + 1$. Assume that one of the following three conditions holds:*

 1) $r = 1$.

 2) $r = 2$ and q square free.

 3) $n \geq 2r$ (instead of $n \geq r + 1$).

Then we have

$$\overline{M} = \bigoplus_S \widetilde{\Theta}(S, q),$$

especially

$$\dim \overline{M} = \sum_S \dim B(S)^{\mathcal{E}(S)} \cdot H(\varrho_0)^{\mathcal{E}(S)}.$$

In principle this is a complete description of all linear relations between the series $\vartheta_{P,\varphi}(S; Z)$. This description involves many quadratic forms. It doesn't give the relations in a single space $\Theta(S, q)$ in a direct manner.

The proof of theorem 1.4 is the content of chapter IV. If f is an arbitrary element of the big space, one considers a Fourier coefficient

$$a(T) \neq 0, \quad T = S[P]; \quad P \text{ primitive}, \quad \text{rank } S = r.$$

(If there is no such Fourier coefficient, we are already done.) We choose T such that $\det S$ is minimal. Then S is a f-kernel form. We investigated in detail the function

$$P \longmapsto a(S[P])$$

in that case. We recall that we proved certain hidden relations. These relations together with the fundemantal lemma allow us to find an element from

$$\tilde{\mathbf{A}}(S) \otimes \mathbf{H}(\varrho_0)^{\mathcal{E}(S)}$$

such that the corresponding theta series has the same Fourier coefficients at all $S[P]$, P primitive (IV sec.5). One substracts this theta series from f and repeats the same procedure if necessary. The number of unimodular classes of f-kernel forms is finite. Hence after a finite number of steps we obtain the desired representation of f as linear combination of theta series.

The big space was a natural home for our theory. We finally want to formulate some consequences for usual modular forms. For this purpose we consider a multiplier system v, which satisfies the assumption I 3.6. We have proved

$$\Gamma_n[q, 2q], \varrho, v] \subset \overline{\mathbf{M}}.$$

On the other hand we have proved that in the case $n \geq r + 2$ all the theta series which generate $\overline{\mathbf{M}}$ are modular forms on $\Gamma_n[q, 2q]$ with respect to certain multiplier systems. Systems of modular forms with pairwise different multiplier systems are of course linearly independent. From 1.4 we obtain

1.5 Theorem. *Assume $n \geq r + 2$. Let S run through a set of representatives of unimodular classes of all positive integral matrices S such that $q^2 S^{-1}$ integral. Assume furthermore that one of the following three conditions holds:*

1) $r = 1$.

2) $r = 2$ *and q square free.*

3) $n \geq 2r$.

Then we have

$$[\Gamma_n[q, 2q], \varrho, v] = \sum_{v=v_S} \Theta(S, q) = \bigoplus_{v=v_S} \tilde{\Theta}(S, q),$$

especially

$$\dim \Gamma_n[q, 2q], \varrho, v] = \sum_{v=v_S} \dim \mathbf{B}(S)^{\mathcal{E}(S)} \cdot \mathbf{H}(\varrho_0)^{\mathcal{E}(S)}.$$

During the proof of the representation theorem it was necessary to get hold of the linear relations between the generating theta series. It is worth while to point out that the result can be improved if one is interested only in a system of generators. The calculations in chap.IV sec.2 actually show that for the generation it is sufficient to restrict to integral forms S such that qS^{-1} (not only $q^2 S^{-1}$) is integral.

1.6 Theorem. *Let S run through a set of representatives of unimodular classes of all positive integral matrices S such that qS^{-1} is half integral if $n = r + 1$ and integral if $n > r + 1$. Assume that one of the following three conditions holds:*

1) $r = 1$.

2) $r = 2$ *and q square free.*

3) $n \geq 2r$ *(instead of* $n \geq r+1$*).*

Then we have

$$\overline{M} = \sum \Theta(S, q);$$

and

$$[\Gamma_n[q, 2q], \varrho, v] = \sum_{v = v_S} \Theta(S, q); \quad if \, n \geq r + 2.$$

Final Remark. The *harmomic* forms P and the *admissible* coefficients φ seem to be tied closely together (1.4). Admissible coefficients should be considered as q-adic harmonic forms. The hidden relations (IV 4.5) can be interpreted as the local equations for pluriharmonicity. The solution of the fundamental lemma is the q-adic analogue of the classification of harmonic forms. We recall that MAASS had restricted to the case $n \geq 2r$ in his classification theorem of scalar valued harmonic forms (II sec.6). KASHIWARA-VERGNE succeeded to overcome this restriction by using the theory of highest weights of rational representations of the orthogonal and general linear group. It is natural to ask, whether their method generalizes to the local case. But the theory of representations of finite linear groups is much more complicated. This is the reason for my impression that the case $r < n < 2r$ is not so easy.

2   An example

In this section we first treat the case

$$r = 3; \quad q = 4$$

and assume that the multiplier system is the third power of the theta multiplier system, i.e. we investigate the space

$$[\Gamma_n[4, 8], 3/2, v_\vartheta^3].$$

This case has been investigated by R. SALVATI MANNI. We closely follow his paper [Man2].

There are well-known examples of modular forms in this space, constructed by means of the thetanullwerte (I sec.5),

$$\vartheta[\mathbf{m}](Z) = \sum_{g \text{ integral}} \exp \pi i \{Z[g + a] + 2b'(g + a)\}.$$

Here

$$\mathbf{m} = \begin{pmatrix} a \\ b \end{pmatrix}; \quad 2a, 2b \in \mathbb{Z}^n \quad \text{(columns)}$$

is the characteristic. Up to the sign these nullwerte depend only on $\mathbf{m} \bmod 1$. Therefore we only consider the finitely many characteristics

$$2\mathbf{m} \in \{0, 1\}^{2n}.$$

The associated theta series do not vanish identically if and only if

$$4a'b = 4 \sum a_\nu b_\nu \equiv 0 \mod 2.$$

Therefore there are

$$(2^n + 1)2^{n-1}$$

different thetanullwerte. These are modular forms of weight $1/2$ and level 4, more precisely

$$\vartheta[\mathbf{m}](M\langle Z\rangle) = v_\vartheta(M)\det(CZ + D)^{1/2}\vartheta[\mathbf{m}](Z) \quad \text{for all } M \in \Gamma_n[4,8].$$

The theta multiplier system takes values 1 and -1 only on $\Gamma_n[4,8]$. Especially v_ϑ^r is trivial for even r.

In this section we prove that in the case $n \geq 5$ each modular form of weight $3/2$ with respect to the group $\Gamma_n[4,8]$ and with respect to the theta multiplier system is a linear combination of three-products of thetanullwerte.

2.1 Proposition. *If $n \geq 5$, the vector space*

$$\left[\Gamma_n[4,8], 3/2, v_\vartheta\right]$$

is generated by the three-products

$$\vartheta[\mathbf{m}_1]\cdots\vartheta[\mathbf{m}_3]; \quad 2\mathbf{m}_\nu \in \{0,1\}^n.$$

Its dimension is

$$\binom{2^{n-1}(2^n + 1) + 2}{3}.$$

The proof rests on our general representation theorem. We have to consider positive definite ternary matrices S such that both

$$S \text{ and } 16S^{-1}$$

are integral. An integral matrix $V \in \mathbb{Z}^{(3,n)}$ is **isotropic**, if the matrix

$$S^{-1}[V + qX]$$

is integral for all integral $(3 \times n)$-matrices X. For a pair (S, V) we consider the theta series

$$\vartheta_V(S; Z) = \sum \exp\frac{\pi i}{q}\sigma\{S[G]Z + 2V'G\}.$$

We know that this is a modular form on $\Gamma_n[4,8]$. The multiplier system is given by

$$v_S\begin{pmatrix} A & B \\ C & D \end{pmatrix} = \mathrm{sgn}(\det D)^{r/2}\left(\frac{(-1)^{r/2}\cdot\det S}{|\det D|}\right), \quad \text{if } \det D \neq 0,$$

where $\left(\frac{\cdot}{\cdot}\right)$ denotes the generalized Legendre-symbol (Jacobi-symbol) (s.I 5.11). This multiplier system is trivial if and only if the determinant of S is a square. Hence our general representation theorem states

2.2 Proposition. *Assume*
$$n \geq 2r.$$
Then $\left[\Gamma_n[4,8], 3/2, v_\vartheta\right]$ *is generated by all*
$$\vartheta_{S,V}; \quad S, qS^{-1} \text{ integral}, \quad \det S \text{ is a square}, \quad V \text{ isotropic}.$$

Remark. *If the set of isotropic V for each S is an additive group, the conditon "$n \geq 2r$" can be replaced by*
$$n \geq r + 2.$$

Simple examples of isotropic matrices are obtained if S can be written in the form
$$S = A'A; \quad A, qA^{-1} \text{ both integral}.$$
Then each matrix $V \in \mathbb{Z}^{(r,n)}$ with the property
$$A^{-1}V \quad \text{integral}$$
is isotropic.

2.3 Definition. *A pair* (S,V)
$$S \text{ integral}, \quad 16S^{-1} \text{ integral}, \quad V \text{ isotropic}$$
is called trivial, if there exists a representation
$$S = A'A$$
with the properties
$$A \text{ integral}, \quad 4A^{-1} \text{ integral}, \quad A^{-1}V \text{ integral}.$$

In [Ma2] Manni shows by means of a certain generalization of the Riemann theta relations due to Mumford (s.[Mu]):

2.4. Remark. *Let* (S,V) *be a trivial pair. Then* $\vartheta_{S,V}$ *is linear combination of products of thetanullwerte*
$$\vartheta[\mathbf{m}_1] \cdots \vartheta[\mathbf{m}_{2r}]; \quad 2\mathbf{m}_\nu \in \{0,1\}^n.$$

Hence proposition 2.1 is a consequence of

2.5. Proposition. *Each pair is trivial. The set of isotropic matrices is always an additive group.*

The proof has been given by a computer calculation.

The dimension formula comes out because the three-products of the thetanullwerte are linearly independent. This means that the case we considered is a very special one. It seems to be reasonable to look for a more general one, where the set of isotropic matrices is not always an additive group and to work out the connection with theta relations. A promising case seems to be the case

$$r = 4; \quad q = 4,$$

i.e. the space

$$[\Gamma_n[4, 8], 2].$$

This space contains the four-products of the thetanullwerte. The famous classical Riemann theta relations are linear relations between them. Generalizing a result of VAN GEEMEN, MANNI [Ma1] proved that each linear relation between the four-products is a consequence of the Riemann relations and he derived a formula for the space generated by all four products.

Using our main result we constructed a system of generators of

$$[\Gamma_n[4, 8], 2] \qquad (n \geq 8).$$

In contrast to the case $r = 3$, it is not always true that the set of isotropic matrices is an additive group. An easy counter example is given by the matrix

$$S = 2E.$$

An integral matrix

$$V = (v_1, \ldots, v_n)$$

is isotropic if and only if the following condition is satisfied

$$v_i' \cdot v_j \equiv 0 \mod 2 \quad (1 \leq i, j \leq n)$$

(especially $v_{i1} + \ldots + v_{ir} \equiv 0 \mod 2$.)

We could not decide whether all theta series belonging to $2E$ are linear combinations of the four-products of thetanullwerte.

If we want to write down an explicit system of generators of

$$[\Gamma_n[4, 8], 2] \qquad (n \geq 8),$$

we first of all need a system of representatives of the unimodular classes of all positive 4×4-matrices S such that S and $16S^{-1}$ are integral and such that $\det S$ is a square. In the appendix to this section we give a complete list of 138 representatives, which has been derived by a computer calculation. In each case it has been verified that S can be represented by the unit matrix, i.e the equation $S = A'A$ has an integral solution with $4A^{-1}$ also integral. But in contrast to the ternary case, the isotropic structures are not always trivial as the above example $(S = 2E)$ shows. For each matrix S in question it is very easy to determine the set of isotropic matrices, but it is difficult to exhibit an explicit basis following the main results (chap.VI sec.1). This demands further non-trivial calculation.

This means that the obvious connection between the theta relations and the isotropic structures has not been completely clarified.

Appendix: A complete system of representatives of all unimodular classes of symmetric positive definite (4×4)-matrices S, such that S and $16 \cdot S^{-1}$ are integral and such that $\det S$ is a square.

$$S_1 = \begin{pmatrix} 1 & 0 & 0 & 0 \\ 0 & 1 & 0 & 0 \\ 0 & 0 & 1 & 0 \\ 0 & 0 & 0 & 1 \end{pmatrix} \qquad S_2 = \begin{pmatrix} 16 & 0 & 0 & 0 \\ 0 & 16 & 0 & 0 \\ 0 & 0 & 16 & 0 \\ 0 & 0 & 0 & 16 \end{pmatrix}$$

$$S_3 = \begin{pmatrix} 1 & 0 & 0 & 0 \\ 0 & 1 & 0 & 0 \\ 0 & 0 & 1 & 0 \\ 0 & 0 & 0 & 4 \end{pmatrix} \qquad S_4 = \begin{pmatrix} 16 & 0 & 0 & 0 \\ 0 & 16 & 0 & 0 \\ 0 & 0 & 16 & 0 \\ 0 & 0 & 0 & 4 \end{pmatrix}$$

$$S_5 = \begin{pmatrix} 1 & 0 & 0 & 0 \\ 0 & 1 & 0 & 0 \\ 0 & 0 & 1 & 0 \\ 0 & 0 & 0 & 16 \end{pmatrix} \qquad S_6 = \begin{pmatrix} 16 & 0 & 0 & 0 \\ 0 & 16 & 0 & 0 \\ 0 & 0 & 16 & 0 \\ 0 & 0 & 0 & 1 \end{pmatrix}$$

$$S_7 = \begin{pmatrix} 1 & 0 & 0 & 0 \\ 0 & 1 & 0 & 0 \\ 0 & 0 & 2 & 0 \\ 0 & 0 & 0 & 2 \end{pmatrix} \qquad S_8 = \begin{pmatrix} 16 & 0 & 0 & 0 \\ 0 & 16 & 0 & 0 \\ 0 & 0 & 8 & 0 \\ 0 & 0 & 0 & 8 \end{pmatrix}$$

$$S_9 = \begin{pmatrix} 1 & 0 & 0 & 0 \\ 0 & 1 & 0 & 0 \\ 0 & 0 & 2 & 0 \\ 0 & 0 & 0 & 8 \end{pmatrix} \qquad S_{10} = \begin{pmatrix} 16 & 0 & 0 & 0 \\ 0 & 16 & 0 & 0 \\ 0 & 0 & 8 & 0 \\ 0 & 0 & 0 & 2 \end{pmatrix}$$

$$S_{11} = \begin{pmatrix} 1 & 0 & 0 & 0 \\ 0 & 1 & 0 & 0 \\ 0 & 0 & 4 & 0 \\ 0 & 0 & 0 & 4 \end{pmatrix} \qquad S_{12} = \begin{pmatrix} 16 & 0 & 0 & 0 \\ 0 & 16 & 0 & 0 \\ 0 & 0 & 4 & 0 \\ 0 & 0 & 0 & 4 \end{pmatrix}$$

$$S_{13} = \begin{pmatrix} 1 & 0 & 0 & 0 \\ 0 & 1 & 0 & 0 \\ 0 & 0 & 4 & 2 \\ 0 & 0 & 2 & 5 \end{pmatrix} \qquad S_{14} = \begin{pmatrix} 16 & 0 & 0 & 0 \\ 0 & 16 & 0 & 0 \\ 0 & 0 & 5 & -2 \\ 0 & 0 & -2 & 4 \end{pmatrix}$$

$$S_{15} = \begin{pmatrix} 1 & 0 & 0 & 0 \\ 0 & 1 & 0 & 0 \\ 0 & 0 & 4 & 0 \\ 0 & 0 & 0 & 16 \end{pmatrix} \qquad S_{16} = \begin{pmatrix} 16 & 0 & 0 & 0 \\ 0 & 16 & 0 & 0 \\ 0 & 0 & 4 & 0 \\ 0 & 0 & 0 & 1 \end{pmatrix}$$

$$S_{17} = \begin{pmatrix} 1 & 0 & 0 & 0 \\ 0 & 1 & 0 & 0 \\ 0 & 0 & 8 & 0 \\ 0 & 0 & 0 & 8 \end{pmatrix} \qquad S_{18} = \begin{pmatrix} 16 & 0 & 0 & 0 \\ 0 & 16 & 0 & 0 \\ 0 & 0 & 2 & 0 \\ 0 & 0 & 0 & 2 \end{pmatrix}$$

$$S_{19} = \begin{pmatrix} 1 & 0 & 0 & 0 \\ 0 & 1 & 0 & 0 \\ 0 & 0 & 16 & 0 \\ 0 & 0 & 0 & 16 \end{pmatrix} \qquad S_{20} = \begin{pmatrix} 1 & 0 & 0 & 0 \\ 0 & 2 & 1 & 1 \\ 0 & 1 & 2 & 1 \\ 0 & 1 & 1 & 2 \end{pmatrix}$$

$$S_{21} = \begin{pmatrix} 16 & 0 & 0 & 0 \\ 0 & 12 & -4 & -4 \\ 0 & -4 & 12 & -4 \\ 0 & -4 & -4 & 12 \end{pmatrix} \qquad S_{22} = \begin{pmatrix} 1 & 0 & 0 & 0 \\ 0 & 2 & 0 & 0 \\ 0 & 0 & 2 & 0 \\ 0 & 0 & 0 & 4 \end{pmatrix}$$

$$S_{23} = \begin{pmatrix} 16 & 0 & 0 & 0 \\ 0 & 8 & 0 & 0 \\ 0 & 0 & 8 & 0 \\ 0 & 0 & 0 & 4 \end{pmatrix} \qquad S_{24} = \begin{pmatrix} 1 & 0 & 0 & 0 \\ 0 & 2 & 0 & -1 \\ 0 & 0 & 2 & 1 \\ 0 & -1 & 1 & 5 \end{pmatrix}$$

$$S_{25} = \begin{pmatrix} 16 & 0 & 0 & 0 \\ 0 & 9 & -1 & 2 \\ 0 & -1 & 9 & -2 \\ 0 & 2 & -2 & 4 \end{pmatrix} \qquad S_{26} = \begin{pmatrix} 1 & 0 & 0 & 0 \\ 0 & 2 & 1 & 1 \\ 0 & 1 & 2 & 1 \\ 0 & 1 & 1 & 6 \end{pmatrix}$$

$$S_{27} = \begin{pmatrix} 16 & 0 & 0 & 0 \\ 0 & 11 & -5 & -1 \\ 0 & -5 & 11 & -1 \\ 0 & -1 & -1 & 3 \end{pmatrix} \qquad S_{28} = \begin{pmatrix} 1 & 0 & 0 & 0 \\ 0 & 2 & 0 & 0 \\ 0 & 0 & 2 & 0 \\ 0 & 0 & 0 & 16 \end{pmatrix}$$

$$S_{29} = \begin{pmatrix} 16 & 0 & 0 & 0 \\ 0 & 8 & 0 & 0 \\ 0 & 0 & 8 & 0 \\ 0 & 0 & 0 & 1 \end{pmatrix} \qquad S_{30} = \begin{pmatrix} 1 & 0 & 0 & 0 \\ 0 & 2 & 0 & 0 \\ 0 & 0 & 3 & 1 \\ 0 & 0 & 1 & 3 \end{pmatrix}$$

$$S_{31} = \begin{pmatrix} 16 & 0 & 0 & 0 \\ 0 & 8 & 0 & 0 \\ 0 & 0 & 6 & -2 \\ 0 & 0 & -2 & 6 \end{pmatrix} \qquad S_{32} = \begin{pmatrix} 1 & 0 & 0 & 0 \\ 0 & 2 & 0 & 0 \\ 0 & 0 & 4 & 0 \\ 0 & 0 & 0 & 8 \end{pmatrix}$$

$$S_{33} = \begin{pmatrix} 16 & 0 & 0 & 0 \\ 0 & 8 & 0 & 0 \\ 0 & 0 & 4 & 0 \\ 0 & 0 & 0 & 2 \end{pmatrix} \qquad S_{34} = \begin{pmatrix} 1 & 0 & 0 & 0 \\ 0 & 2 & 0 & 0 \\ 0 & 0 & 6 & 2 \\ 0 & 0 & 2 & 6 \end{pmatrix}$$

$$S_{35} = \begin{pmatrix} 16 & 0 & 0 & 0 \\ 0 & 8 & 0 & 0 \\ 0 & 0 & 3 & -1 \\ 0 & 0 & -1 & 3 \end{pmatrix} \qquad S_{36} = \begin{pmatrix} 1 & 0 & 0 & 0 \\ 0 & 2 & 0 & 0 \\ 0 & 0 & 8 & 0 \\ 0 & 0 & 0 & 16 \end{pmatrix}$$

$$S_{37} = \begin{pmatrix} 1 & 0 & 0 & 0 \\ 0 & 3 & 1 & -1 \\ 0 & 1 & 3 & 1 \\ 0 & -1 & 1 & 3 \end{pmatrix} \qquad S_{38} = \begin{pmatrix} 16 & 0 & 0 & 0 \\ 0 & 8 & -4 & 4 \\ 0 & -4 & 8 & -4 \\ 0 & 4 & -4 & 8 \end{pmatrix}$$

$$S_{39} = \begin{pmatrix} 1 & 0 & 0 & 0 \\ 0 & 3 & 1 & 0 \\ 0 & 1 & 3 & 0 \\ 0 & 0 & 0 & 8 \end{pmatrix} \qquad S_{40} = \begin{pmatrix} 16 & 0 & 0 & 0 \\ 0 & 6 & -2 & 0 \\ 0 & -2 & 6 & 0 \\ 0 & 0 & 0 & 2 \end{pmatrix}$$

$$S_{41} = \begin{pmatrix} 1 & 0 & 0 & 0 \\ 0 & 3 & 1 & -1 \\ 0 & 1 & 11 & 5 \\ 0 & -1 & 5 & 11 \end{pmatrix} \qquad S_{42} = \begin{pmatrix} 1 & 0 & 0 & 0 \\ 0 & 4 & 0 & 0 \\ 0 & 0 & 4 & 0 \\ 0 & 0 & 0 & 4 \end{pmatrix}$$

$$S_{43} = \begin{pmatrix} 16 & 0 & 0 & 0 \\ 0 & 4 & 0 & 0 \\ 0 & 0 & 4 & 0 \\ 0 & 0 & 0 & 4 \end{pmatrix} \qquad S_{44} = \begin{pmatrix} 1 & 0 & 0 & 0 \\ 0 & 4 & 0 & -2 \\ 0 & 0 & 4 & 0 \\ 0 & -2 & 0 & 5 \end{pmatrix}$$

$$S_{45} = \begin{pmatrix} 16 & 0 & 0 & 0 \\ 0 & 5 & 0 & 2 \\ 0 & 0 & 4 & 0 \\ 0 & 2 & 0 & 4 \end{pmatrix} \qquad S_{46} = \begin{pmatrix} 1 & 0 & 0 & 0 \\ 0 & 4 & 0 & -2 \\ 0 & 0 & 4 & 2 \\ 0 & -2 & 2 & 6 \end{pmatrix}$$

$$S_{47} = \begin{pmatrix} 16 & 0 & 0 & 0 \\ 0 & 5 & -1 & 2 \\ 0 & -1 & 5 & -2 \\ 0 & 2 & -2 & 4 \end{pmatrix} \qquad S_{48} = \begin{pmatrix} 1 & 0 & 0 & 0 \\ 0 & 4 & 0 & 0 \\ 0 & 0 & 4 & 0 \\ 0 & 0 & 0 & 16 \end{pmatrix}$$

$$S_{49} = \begin{pmatrix} 1 & 0 & 0 & 0 \\ 0 & 4 & 2 & 2 \\ 0 & 2 & 5 & 1 \\ 0 & 2 & 1 & 5 \end{pmatrix} \qquad S_{50} = \begin{pmatrix} 16 & 0 & 0 & 0 \\ 0 & 6 & -2 & -2 \\ 0 & -2 & 4 & 0 \\ 0 & -2 & 0 & 4 \end{pmatrix}$$

$$S_{51} = \begin{pmatrix} 1 & 0 & 0 & 0 \\ 0 & 4 & 2 & 0 \\ 0 & 2 & 5 & 0 \\ 0 & 0 & 0 & 16 \end{pmatrix} \qquad S_{52} = \begin{pmatrix} 1 & 0 & 0 & 0 \\ 0 & 4 & 0 & 0 \\ 0 & 0 & 8 & 0 \\ 0 & 0 & 0 & 8 \end{pmatrix}$$

$$S_{53} = \begin{pmatrix} 1 & 0 & 0 & 0 \\ 0 & 4 & 2 & 2 \\ 0 & 2 & 9 & 1 \\ 0 & 2 & 1 & 9 \end{pmatrix} \qquad S_{54} = \begin{pmatrix} 1 & 0 & 0 & 0 \\ 0 & 6 & 2 & 0 \\ 0 & 2 & 6 & 0 \\ 0 & 0 & 0 & 8 \end{pmatrix}$$

$$S_{55} = \begin{pmatrix} 1 & 0 & 0 & 0 \\ 0 & 8 & 4 & -4 \\ 0 & 4 & 8 & 0 \\ 0 & -4 & 0 & 8 \end{pmatrix} \qquad S_{56} = \begin{pmatrix} 2 & 0 & 1 & 1 \\ 0 & 2 & 1 & -1 \\ 1 & 1 & 2 & 0 \\ 1 & -1 & 0 & 2 \end{pmatrix}$$

$$S_{57} = \begin{pmatrix} 16 & 0 & -8 & -8 \\ 0 & 16 & -8 & 8 \\ -8 & -8 & 16 & 0 \\ -8 & 8 & 0 & 16 \end{pmatrix} \qquad S_{58} = \begin{pmatrix} 2 & 0 & 0 & 0 \\ 0 & 2 & 0 & 0 \\ 0 & 0 & 2 & 0 \\ 0 & 0 & 0 & 2 \end{pmatrix}$$

$$S_{59} = \begin{pmatrix} 8 & 0 & 0 & 0 \\ 0 & 8 & 0 & 0 \\ 0 & 0 & 8 & 0 \\ 0 & 0 & 0 & 8 \end{pmatrix} \qquad S_{60} = \begin{pmatrix} 2 & 0 & 0 & -1 \\ 0 & 2 & 0 & -1 \\ 0 & 0 & 2 & 0 \\ -1 & -1 & 0 & 3 \end{pmatrix}$$

$$S_{61} = \begin{pmatrix} 10 & 2 & 0 & 4 \\ 2 & 10 & 0 & 4 \\ 0 & 0 & 8 & 0 \\ 4 & 4 & 0 & 8 \end{pmatrix} \qquad S_{62} = \begin{pmatrix} 2 & 1 & 1 & 0 \\ 1 & 2 & 1 & 0 \\ 1 & 1 & 2 & 0 \\ 0 & 0 & 0 & 4 \end{pmatrix}$$

$$S_{63} = \begin{pmatrix} 12 & -4 & -4 & 0 \\ -4 & 12 & -4 & 0 \\ -4 & -4 & 12 & 0 \\ 0 & 0 & 0 & 4 \end{pmatrix} \qquad S_{64} = \begin{pmatrix} 2 & 1 & 1 & -1 \\ 1 & 2 & 1 & -1 \\ 1 & 1 & 2 & 0 \\ -1 & -1 & 0 & 5 \end{pmatrix}$$

$$S_{65} = \begin{pmatrix} 13 & -3 & -5 & 2 \\ -3 & 13 & -5 & 2 \\ -5 & -5 & 13 & -2 \\ 2 & 2 & -2 & 4 \end{pmatrix} \qquad S_{66} = \begin{pmatrix} 2 & 0 & 0 & 0 \\ 0 & 2 & 0 & 0 \\ 0 & 0 & 2 & 0 \\ 0 & 0 & 0 & 8 \end{pmatrix}$$

$$S_{67} = \begin{pmatrix} 8 & 0 & 0 & 0 \\ 0 & 8 & 0 & 0 \\ 0 & 0 & 8 & 0 \\ 0 & 0 & 0 & 2 \end{pmatrix} \qquad S_{68} = \begin{pmatrix} 2 & 1 & 1 & 0 \\ 1 & 2 & 1 & 0 \\ 1 & 1 & 2 & 0 \\ 0 & 0 & 0 & 16 \end{pmatrix}$$

$$S_{69} = \begin{pmatrix} 12 & -4 & -4 & 0 \\ -4 & 12 & -4 & 0 \\ -4 & -4 & 12 & 0 \\ 0 & 0 & 0 & 1 \end{pmatrix} \qquad S_{70} = \begin{pmatrix} 2 & 1 & 1 & 1 \\ 1 & 2 & 1 & 1 \\ 1 & 1 & 3 & 1 \\ 1 & 1 & 1 & 3 \end{pmatrix}$$

$$S_{71} = \begin{pmatrix} 12 & -4 & -2 & -2 \\ -4 & 12 & -2 & -2 \\ -2 & -2 & 7 & -1 \\ -2 & -2 & -1 & 7 \end{pmatrix} \qquad S_{72} = \begin{pmatrix} 2 & 0 & -1 & -1 \\ 0 & 2 & 1 & -1 \\ -1 & 1 & 3 & 0 \\ -1 & -1 & 0 & 3 \end{pmatrix}$$

$$S_{73} = \begin{pmatrix} 12 & 0 & 4 & 4 \\ 0 & 12 & -4 & 4 \\ 4 & -4 & 8 & 0 \\ 4 & 4 & 0 & 8 \end{pmatrix} \qquad S_{74} = \begin{pmatrix} 2 & 0 & -1 & 0 \\ 0 & 2 & 1 & 0 \\ -1 & 1 & 3 & 0 \\ 0 & 0 & 0 & 8 \end{pmatrix}$$

$$S_{75} = \begin{pmatrix} 10 & -2 & 4 & 0 \\ -2 & 10 & -4 & 0 \\ 4 & -4 & 8 & 0 \\ 0 & 0 & 0 & 2 \end{pmatrix} \qquad S_{76} = \begin{pmatrix} 2 & 0 & 0 & 0 \\ 0 & 2 & 0 & 0 \\ 0 & 0 & 4 & 0 \\ 0 & 0 & 0 & 4 \end{pmatrix}$$

$$S_{77} = \begin{pmatrix} 8 & 0 & 0 & 0 \\ 0 & 8 & 0 & 0 \\ 0 & 0 & 4 & 0 \\ 0 & 0 & 0 & 4 \end{pmatrix} \qquad S_{78} = \begin{pmatrix} 2 & 0 & 0 & -1 \\ 0 & 2 & 0 & -1 \\ 0 & 0 & 4 & 0 \\ -1 & -1 & 0 & 5 \end{pmatrix}$$

$$S_{79} = \begin{pmatrix} 9 & 1 & 0 & 2 \\ 1 & 9 & 0 & 2 \\ 0 & 0 & 4 & 0 \\ 2 & 2 & 0 & 4 \end{pmatrix} \qquad S_{80} = \begin{pmatrix} 2 & 0 & 0 & 0 \\ 0 & 2 & 0 & 0 \\ 0 & 0 & 4 & 2 \\ 0 & 0 & 2 & 5 \end{pmatrix}$$

$$S_{81} = \begin{pmatrix} 8 & 0 & 0 & 0 \\ 0 & 8 & 0 & 0 \\ 0 & 0 & 5 & -2 \\ 0 & 0 & -2 & 4 \end{pmatrix} \qquad S_{82} = \begin{pmatrix} 2 & 1 & 0 & -1 \\ 1 & 2 & 0 & -1 \\ 0 & 0 & 4 & 0 \\ -1 & -1 & 0 & 6 \end{pmatrix}$$

$$S_{83} = \begin{pmatrix} 11 & -5 & 0 & 1 \\ -5 & 11 & 0 & 1 \\ 0 & 0 & 4 & 0 \\ 1 & 1 & 0 & 3 \end{pmatrix} \qquad S_{84} = \begin{pmatrix} 2 & 0 & 0 & -1 \\ 0 & 2 & 0 & -1 \\ 0 & 0 & 4 & 2 \\ -1 & -1 & 2 & 6 \end{pmatrix}$$

$$S_{85} = \begin{pmatrix} 9 & 1 & -1 & 2 \\ 1 & 9 & -1 & 2 \\ -1 & -1 & 5 & -2 \\ 2 & 2 & -2 & 4 \end{pmatrix} \qquad S_{86} = \begin{pmatrix} 2 & 0 & 0 & 0 \\ 0 & 2 & 0 & 0 \\ 0 & 0 & 4 & 0 \\ 0 & 0 & 0 & 16 \end{pmatrix}$$

$$S_{87} = \begin{pmatrix} 2 & 0 & -1 & -1 \\ 0 & 2 & 1 & -1 \\ -1 & 1 & 5 & 0 \\ -1 & -1 & 0 & 5 \end{pmatrix} \qquad S_{88} = \begin{pmatrix} 10 & 0 & 2 & 2 \\ 0 & 10 & -2 & 2 \\ 2 & -2 & 4 & 0 \\ 2 & 2 & 0 & 4 \end{pmatrix}$$

$$S_{89} = \begin{pmatrix} 2 & 0 & -1 & 0 \\ 0 & 2 & 1 & 0 \\ -1 & 1 & 5 & 0 \\ 0 & 0 & 0 & 16 \end{pmatrix} \qquad S_{90} = \begin{pmatrix} 2 & 1 & 1 & -1 \\ 1 & 2 & 1 & -1 \\ 1 & 1 & 6 & 2 \\ -1 & -1 & 2 & 6 \end{pmatrix}$$

$$S_{91} = \begin{pmatrix} 12 & -4 & -2 & 2 \\ -4 & 12 & -2 & 2 \\ -2 & -2 & 4 & -2 \\ 2 & 2 & -2 & 4 \end{pmatrix} \qquad S_{92} = \begin{pmatrix} 2 & 1 & 1 & 0 \\ 1 & 2 & 1 & 0 \\ 1 & 1 & 6 & 0 \\ 0 & 0 & 0 & 16 \end{pmatrix}$$

$$S_{93} = \begin{pmatrix} 2 & 0 & 0 & 0 \\ 0 & 2 & 0 & 0 \\ 0 & 0 & 8 & 0 \\ 0 & 0 & 0 & 8 \end{pmatrix} \qquad S_{94} = \begin{pmatrix} 2 & 0 & -1 & -1 \\ 0 & 2 & 1 & -1 \\ -1 & 1 & 9 & 0 \\ -1 & -1 & 0 & 9 \end{pmatrix}$$

$$S_{95} = \begin{pmatrix} 2 & 0 & 0 & 0 \\ 0 & 3 & 1 & 0 \\ 0 & 1 & 3 & 0 \\ 0 & 0 & 0 & 4 \end{pmatrix} \qquad S_{96} = \begin{pmatrix} 8 & 0 & 0 & 0 \\ 0 & 6 & -2 & 0 \\ 0 & -2 & 6 & 0 \\ 0 & 0 & 0 & 4 \end{pmatrix}$$

$$S_{97} = \begin{pmatrix} 2 & 0 & 0 & 0 \\ 0 & 3 & 1 & -1 \\ 0 & 1 & 3 & 1 \\ 0 & -1 & 1 & 5 \end{pmatrix} \qquad S_{98} = \begin{pmatrix} 8 & 0 & 0 & 0 \\ 0 & 7 & -3 & 2 \\ 0 & -3 & 7 & -2 \\ 0 & 2 & -2 & 4 \end{pmatrix}$$

$$S_{99} = \begin{pmatrix} 2 & 1 & 1 & 0 \\ 1 & 3 & 1 & 1 \\ 1 & 1 & 3 & 1 \\ 0 & 1 & 1 & 6 \end{pmatrix} \qquad S_{100} = \begin{pmatrix} 11 & -3 & -3 & 1 \\ -3 & 7 & -1 & -1 \\ -3 & -1 & 7 & -1 \\ 1 & -1 & -1 & 3 \end{pmatrix}$$

$$S_{101} = \begin{pmatrix} 2 & 0 & 0 & 0 \\ 0 & 3 & 1 & 0 \\ 0 & 1 & 3 & 0 \\ 0 & 0 & 0 & 16 \end{pmatrix} \qquad S_{102} = \begin{pmatrix} 2 & 1 & 1 & 0 \\ 1 & 3 & 1 & -1 \\ 1 & 1 & 7 & 3 \\ 0 & -1 & 3 & 10 \end{pmatrix}$$

$$S_{103} = \begin{pmatrix} 2 & 0 & 0 & 0 \\ 0 & 4 & 2 & 2 \\ 0 & 2 & 4 & 2 \\ 0 & 2 & 2 & 4 \end{pmatrix} \qquad S_{104} = \begin{pmatrix} 8 & 0 & 0 & 0 \\ 0 & 6 & -2 & -2 \\ 0 & -2 & 6 & -2 \\ 0 & -2 & -2 & 6 \end{pmatrix}$$

$$S_{105} = \begin{pmatrix} 2 & 0 & 0 & 0 \\ 0 & 4 & 0 & 0 \\ 0 & 0 & 4 & 0 \\ 0 & 0 & 0 & 8 \end{pmatrix} \qquad S_{106} = \begin{pmatrix} 2 & 0 & 0 & 0 \\ 0 & 4 & 2 & 0 \\ 0 & 2 & 5 & 0 \\ 0 & 0 & 0 & 8 \end{pmatrix}$$

$$S_{107} = \begin{pmatrix} 2 & 0 & 0 & 0 \\ 0 & 4 & 0 & 0 \\ 0 & 0 & 6 & 2 \\ 0 & 0 & 2 & 6 \end{pmatrix} \qquad S_{108} = \begin{pmatrix} 2 & 0 & 0 & 0 \\ 0 & 4 & 2 & 2 \\ 0 & 2 & 7 & 3 \\ 0 & 2 & 3 & 7 \end{pmatrix}$$

$$S_{109} = \begin{pmatrix} 2 & 1 & 0 & -1 \\ 1 & 6 & 1 & -3 \\ 0 & 1 & 6 & 1 \\ -1 & -3 & 1 & 6 \end{pmatrix} \qquad S_{110} = \begin{pmatrix} 2 & 0 & 0 & 0 \\ 0 & 6 & 2 & -2 \\ 0 & 2 & 6 & 2 \\ 0 & -2 & 2 & 6 \end{pmatrix}$$

$$S_{111} = \begin{pmatrix} 3 & 1 & 0 & 0 \\ 1 & 3 & 0 & 0 \\ 0 & 0 & 3 & 1 \\ 0 & 0 & 1 & 3 \end{pmatrix} \qquad S_{112} = \begin{pmatrix} 6 & -2 & 0 & 0 \\ -2 & 6 & 0 & 0 \\ 0 & 0 & 6 & -2 \\ 0 & 0 & -2 & 6 \end{pmatrix}$$

$$S_{113} = \begin{pmatrix} 3 & 1 & -1 & 0 \\ 1 & 3 & 1 & 0 \\ -1 & 1 & 3 & 0 \\ 0 & 0 & 0 & 4 \end{pmatrix} \qquad S_{114} = \begin{pmatrix} 8 & -4 & 4 & 0 \\ -4 & 8 & -4 & 0 \\ 4 & -4 & 8 & 0 \\ 0 & 0 & 0 & 4 \end{pmatrix}$$

$$S_{115} = \begin{pmatrix} 3 & 1 & -1 & -1 \\ 1 & 3 & 1 & -1 \\ -1 & 1 & 3 & 1 \\ -1 & -1 & 1 & 5 \end{pmatrix} \qquad S_{116} = \begin{pmatrix} 8 & -4 & 4 & 0 \\ -4 & 9 & -5 & 2 \\ 4 & -5 & 9 & -2 \\ 0 & 2 & -2 & 4 \end{pmatrix}$$

$$S_{117} = \begin{pmatrix} 3 & 1 & -1 & 0 \\ 1 & 3 & 1 & 0 \\ -1 & 1 & 3 & 0 \\ 0 & 0 & 0 & 16 \end{pmatrix} \qquad S_{118} = \begin{pmatrix} 3 & 1 & -1 & -1 \\ 1 & 3 & 1 & 1 \\ -1 & 1 & 4 & 2 \\ -1 & 1 & 2 & 4 \end{pmatrix}$$

$$S_{119} = \begin{pmatrix} 8 & -4 & 2 & 2 \\ -4 & 8 & -2 & -2 \\ 2 & -2 & 6 & -2 \\ 2 & -2 & -2 & 6 \end{pmatrix} \qquad S_{120} = \begin{pmatrix} 3 & 1 & 0 & 0 \\ 1 & 3 & 0 & 0 \\ 0 & 0 & 4 & 0 \\ 0 & 0 & 0 & 8 \end{pmatrix}$$

$$S_{121} = \begin{pmatrix} 3 & 1 & -1 & 0 \\ 1 & 3 & 1 & 0 \\ -1 & 1 & 5 & 0 \\ 0 & 0 & 0 & 8 \end{pmatrix} \qquad S_{122} = \begin{pmatrix} 3 & 0 & -1 & -1 \\ 0 & 3 & 1 & -1 \\ -1 & 1 & 6 & 0 \\ -1 & -1 & 0 & 6 \end{pmatrix}$$

$$S_{123} = \begin{pmatrix} 3 & 1 & 0 & 0 \\ 1 & 3 & 0 & 0 \\ 0 & 0 & 6 & 2 \\ 0 & 0 & 2 & 6 \end{pmatrix} \qquad S_{124} = \begin{pmatrix} 3 & 1 & -1 & -1 \\ 1 & 3 & 1 & 1 \\ -1 & 1 & 7 & 3 \\ -1 & 1 & 3 & 7 \end{pmatrix}$$

$$S_{125} = \begin{pmatrix} 3 & 1 & 1 & 0 \\ 1 & 4 & 2 & -1 \\ 1 & 2 & 4 & 1 \\ 0 & -1 & 1 & 9 \end{pmatrix} \qquad S_{126} = \begin{pmatrix} 4 & 0 & 2 & 2 \\ 0 & 4 & 2 & -2 \\ 2 & 2 & 4 & 0 \\ 2 & -2 & 0 & 4 \end{pmatrix}$$

$$S_{127} = \begin{pmatrix} 8 & 0 & -4 & -4 \\ 0 & 8 & -4 & 4 \\ -4 & -4 & 8 & 0 \\ -4 & 4 & 0 & 8 \end{pmatrix} \qquad S_{128} = \begin{pmatrix} 4 & 0 & 0 & 0 \\ 0 & 4 & 0 & 0 \\ 0 & 0 & 4 & 0 \\ 0 & 0 & 0 & 4 \end{pmatrix}$$

$$S_{129} = \begin{pmatrix} 4 & 0 & 0 & 0 \\ 0 & 4 & 0 & -2 \\ 0 & 0 & 4 & 0 \\ 0 & -2 & 0 & 5 \end{pmatrix} \qquad S_{130} = \begin{pmatrix} 4 & 0 & 0 & -2 \\ 0 & 4 & 0 & -2 \\ 0 & 0 & 4 & 0 \\ -2 & -2 & 0 & 6 \end{pmatrix}$$

$$S_{131} = \begin{pmatrix} 4 & 0 & 0 & -2 \\ 0 & 4 & 0 & -2 \\ 0 & 0 & 4 & 2 \\ -2 & -2 & 2 & 7 \end{pmatrix} \qquad S_{132} = \begin{pmatrix} 4 & 2 & 2 & 0 \\ 2 & 4 & 2 & 0 \\ 2 & 2 & 4 & 0 \\ 0 & 0 & 0 & 8 \end{pmatrix}$$

$$S_{133} = \begin{pmatrix} 4 & 0 & -2 & 0 \\ 0 & 4 & 0 & -2 \\ -2 & 0 & 5 & 0 \\ 0 & -2 & 0 & 5 \end{pmatrix} \qquad S_{134} = \begin{pmatrix} 4 & 0 & -2 & -2 \\ 0 & 4 & 0 & 0 \\ -2 & 0 & 5 & 1 \\ -2 & 0 & 1 & 5 \end{pmatrix}$$

$$S_{135} = \begin{pmatrix} 4 & 0 & -2 & -2 \\ 0 & 4 & 0 & -2 \\ -2 & 0 & 5 & 1 \\ -2 & -2 & 1 & 6 \end{pmatrix} \qquad S_{136} = \begin{pmatrix} 4 & 0 & -2 & -2 \\ 0 & 4 & 2 & -2 \\ -2 & 2 & 6 & 0 \\ -2 & -2 & 0 & 6 \end{pmatrix}$$

$$S_{137} = \begin{pmatrix} 4 & 2 & 2 & 2 \\ 2 & 5 & 1 & 1 \\ 2 & 1 & 5 & 1 \\ 2 & 1 & 1 & 5 \end{pmatrix} \qquad S_{138} = \begin{pmatrix} 4 & 2 & 2 & 0 \\ 2 & 5 & 1 & -2 \\ 2 & 1 & 5 & 2 \\ 0 & -2 & 2 & 6 \end{pmatrix}$$

References

[AM] Andrianov,A.N., Maloletkin,G.N.: *Behaviour of theta series of degree N under modular substitutions*, Math. USSR Izvestija **9**, 227-241 (1975)

[Al] Althoff,B.: *Darstellung singulärer Modulformen zur Gruppe $\Gamma_0^{(n)}[q]$ als Linearkombination von Thetareihen mit harmonischen Koeffizienten*, Diplomarbeit, Heidelberg (1985)

[Ei] Eichler,M.: *Einführung in die Theorie der algebraischen Zahlen und Funktionen*, Basel-Stutttgart: Birkhäuser (1963)

[EZ] Eichler,Z. Zagier,D.: *The theory of Jacobi forms*, Progress in Mathematics, Vol. **55**. Boston-Basel-Stuttgart: Birkhäuser (1985)

[En] Endres,R.: *Über die Darstellung singulärer Modulformen halbzahligen Gewichts durch Thetareihen*, Math.Z. **193**, 15-40 (1986)

[Fr1] Freitag,E.: *Holomorphe Differentialformen zu Kongruenzgruppen der Siegelschen Modulgruppe zweiten Grades*, Math.Ann. **216**, 155-164 (1975)

[Fr2] Freitag,E.: *Holomorphe Differentialformen zu Kongruenzgruppen der Siegelschen Modulgruppe*, Inv. math. **30**, 181-196 (1975)

[Fr3] Freitag,E.: *The transformation formalism of vector valued theta functions with respect to the Siegel modular group*, Journal of the Indian Math. Soc. **52**, 185-207 (1987)

[Fr4] Freitag,E.: *Siegelsche Modulformen*, Grundlehren der mathematischen Wissenschaften, Bd. **254**. Berlin-Heidelberg-New York: Springer (1983)

[Fr5] Freitag,E.: *Thetareihen mit harmonischen Koeffizienten zur Siegelschen Modulgruppe*, Math. Ann. **254**, 27-51 (1980)

[Fr6] Freitag,E.: *Ein Verschwindungssatz für automorphe Formen zur Siegelschen Modulgruppe* Math. Z. **165**, 11-18 (1979)

[Fr7] Freitag,E.: *Singular modular forms*, Forschungsschwerpunkt Geometrie, Heidelberg, Nr.24 (1988)

[Fr8] Freitag,E.: *Ein kombinatorisches Lemma I–IV*, Forschungsschwerpunkt Geometrie, Heidelberg, Nr.21,22,23,30 (1988)

[Fr9] Freitag,E.: *Degenerierende Formen in Matrixvariablen über endlichen Körpern*, Forschungsschwerpunkt Geometrie, Heidelberg, Nr.41 (1988)

[Fr10] Freitag,E.: *Degenerierende Formen in Matrixvariablen über endlichen Ringen*, Forschungsschwerpunkt Geometrie, Heidelberg, Nr.45 (1989)

[Ha] Hasse,H.: *Vorlesung über Zahlentheorie*, Berlin-Göttingen-Heidelberg: Springer (1950)

[Ho] Howe,R.: *Automorphic forms of low rank*, in: Non commutative harmonic analysis and Lie groups. Lectures in Math. **880**. Berlin-Heidelberg-New York: Springer 1981

[Ig1] Igusa,J.I.: *On the graded ring of theta constants*, Amer. J. Math. **86**, 219-246 (1964)

[Ig2] Igusa,J.I.: *On the graded ring of theta constants II*, Amer. J. Math. **88**, 221-236 (1966)

[Ig3] *Theta functions*, Grundlehren der mathematischen Wissenschaften, Bd. **194**. Berlin-Heidelberg-New York: Springer (1972)

[KV] Kashiwara,M., Vergne,M.: *On the Segal-Shale-Weil representation and harmonic polynomials*, Invent.Math. **44**, 1-47 (1978)

[KW] Krazer,A., Wirtinger,W.: *Abelsche Funktionen und allgemeine Thetafunktionen*. Enzykl.Math.Wiss., II B7, 604-873

[an1] Manni,R.S.: *On the dimension of the space* $\mathbb{C}[\theta_m]_4$, Nagoya Math. J. Vol. **98**, 99-107 (1985)

[an2] Manni,R.S.: *Singular modular forms of weight 3/2 and thetanullwerte*. Forschungsschwerpunkt Geometrie, Heidelberg, Nr.40 (1988)

[Ma1] Maaß,H.: *Siegel's modular forms and Dirichlet series*. Lecture Notes in Math. **216** Springer Verlag Berlin-Heidelberg-New York (1971)

[Ma2] Maaß,H.: *Harmonische Formen in einer Matrixvariablen*. Math.Ann. **252**, 133-140 (1980)

[Me] Mennicke,J.: *Zur Theorie der Siegelschen Modulgruppe* Math. Ann. **159**, 115-129 (1965)

[Mu] Mumford,D.: *Tata Lectures on Theta I*, Progress in Mathematics, Vol. **28**. Boston-Basel-Stuttgart: Birkhäuser (1983)

[NS] Naimark,M.A. Stern,A.I.: *Theory of group representations* Grundlehren der mathematischen Wissenschaften, Bd. **246**. Berlin-Heidelberg-New York: Springer (1982)

[Re1] Resnikoff,H.L.: *On a class of linear differential equations for automorphic forms in several complex variables*, Am. J. Math. **95**, 321-332 (1973)

[Re2] Resnikoff,H.L.: *Automorphic forms of singular weight are singular forms*, Math. Ann. **215**, 173-193 (1975)

[Sh] Shimura,G.: *On certain reciprocity laws for theta functions and modular forms*, Acta Math. **141**, 35-71 (1978)

[Sta] Stark,H.M.: *On the transformation formula for the symplectic theta function and applications*, J. Fac. Sci. Univ. Tokyo Sect. I A Math. **29**, 1-12 (1982)

[Sty1] Styer,R.: *Prime determinant matrices and the symplectic theta function*, Am. J. Math. **106**, 645-664 (1984)

[Sty2] Styer,R.: *Evaluating symplectic Gauss sums and Jacobi symbols*, Nagoya Math. J. **95**, 1-23 (1984)

[We1] Weissauer,R.: *Vektorwertige Modulformen kleinen Gewichts*, J. f. reine u. angew. Math. **343**, 184-202 (1983)

[We2] Weissauer,R.: *Stabile Modulformen und Eisensteinreihen*, Lectures in Math. 1219. Berlin-Heidelberg-New York: Springer 1986

[Zi] Ziegler,C.: *Jacobi forms of higher degree*, Hamb.Abh. **53**, (1989)

[Zw] Zwecker, E.: *Ein Zusammenhang zwischen Klassenzahlen imaginär- quadratischer Zahlkörper und Invarianten gewisser Körper von Siegelschen Modulfunktionen beliebigen Grades*, Diplomarbeit, Heidelberg (1985)